ABOUT THE EDITOR

JOHN BROCKMAN is the author/editor of twenty books. He is the founder of Brockman, Inc., a literary and software agency, founder of The Reality Club, president of Edge Foundation, Inc., and editor and publisher of *Edge*, a Web site (www.edge.org) forum for leading scientists and thinkers.

BOOKS BY JOHN BROCKMAN

As Author:

By the Late John Brockman
37
Afterwords
The Third Culture: Beyond the Scientific Revolution
Digerati

As Editor:

About Bateson
Speculations
Doing Science
Ways of Knowing
Creativity
The Greatest Inventions of the Past 2,000 Years

As Coeditor:

How Things Are

THE
NEXT
FIFTY
YEARS

THE
NEXT
FIFTY
YEARS

❏

SCIENCE IN THE FIRST HALF
OF THE TWENTY-FIRST CENTURY

❏

Edited by John Brockman

VINTAGE BOOKS
A Division of Random House, Inc.
New York

A VINTAGE ORIGINAL, MAY 2002

Copyright © 2002 by John Brockman

All rights reserved under International and
Pan-American Copyright Conventions. Published in the United States by
Vintage Books, a division of Random House, Inc., New York,
and simultaneously in Canada by Random House
of Canada Limited, Toronto.

Vintage Books and colophon are
registered trademarks of Random House, Inc.

Library of Congress Cataloging-in-Publication Data

The next fifty years : science in the first half of the twenty-first century /
edited by John Brockman.
p. cm.
ISBN 0-375-71342-5 (trade paper)
1. Science—Forecasting.   I. Brockman, John, 1941–

Q125 .N485 2002
501'.12—dc21
2001057368

Book design by Oksana Kushnir

www.vintagebooks.com

Printed in the United States of America
10   9   8   7   6   5   4   3   2

To my son,
Max Brockman

CONTENTS

ACKNOWLEDGMENTS

Marty Asher, publisher of Vintage Books, suggested that a collection of original essays by scientists on the next fifty years would make a valuable book, and I thank him for the suggestion and his encouragement. I am also indebted to Sara Lippincott for her thoughtful and meticulous editing.

In 1991 I published an essay entitled "The Third Culture," in which I introduced the idea that a new culture, a public culture, had come into being which consisted of "those scientists and other thinkers in the empirical world who, through their work and expository writing, are taking the place of the traditional intellectual in rendering visible the deeper meanings of our lives, redefining who and what we are."

Science is the big news, and it is scientists who are asking the big questions. Through their books and articles they have become the new public intellectuals, leaders of a new kind of public culture. *The Next Fifty Years* pictures several aspects of this new culture.

The essays presented here are not the marginal discussions of the old-style intellectual culture; the work of this group of scientists centers on developments that affect the lives of everybody on the planet. Consider the recent international press coverage of such issues as stem-cell research, cloning, the sequencing of the human genome, artificial intelligence, astrobiology, and quantum computing. The topics (as well as the work) inevitably cut across disciplines. One of the reasons for the remarkable increase over the last ten years in readership of books by scientists (including those who are contributors to this volume) is that they are compelled to write in language that their colleagues in other disciplines can understand. The generally educated reader thus benefits, because he or she can now join in and look over the shoulders of members of this group as they take on the big questions of the day.

In this culture, and in the present book, scientists are not writing popularizations meant only to entertain the public; they are writing for, and engaging, their peers in other disciplines in the debates of our times. The goal is not the popularization of science but the attempt to make the latest scientific research understandable *within science itself* as well as to a wide audience.

That said, I do not claim that the authors of these essays necessarily offer "better" answers to the questions that arise in our daily lives than does any average person. The critical difference is in the quality of the questions they address.

The subject, and a starting point for these twenty-five original essays, is "the next fifty years" in the respective fields of the contributors. How will achievements in science over the next half century change our world? How will they change the questions we are asking about who and what we are? What developments might we expect in each field or discipline, and how might these influence and cut across other disciplines? What current expectations will not be realized, and what will be the surprising shifts in perception?

The book features thoughtful, challenging essays—intellectual adventures—by twenty-five leading scientists, all of them frequent communicators of their science in books and articles for the general public. They are the biologists Richard Dawkins, Paul W. Ewald, Brian Goodwin, Stuart Kauffman, and Robert Sapolsky; the chemist Peter Atkins; the psychologists Paul Bloom, Mihaly Csikszentmihalyi, Nancy Etcoff, Alison Gopnik, Judith Rich Harris, and Geoffrey Miller; the psychologist and computer scientist John H. Holland; the psychologist and AI researcher Roger C. Schank; neuroscientists Samuel Barondes, Marc D. Hauser, and Joseph LeDoux; computer scientists David Gelernter and Jaron Lanier; Rodney Brooks, director of MIT's Artificial Intelligence Laboratory; the mathematicians Ian Stewart and Steven Strogatz; the astronomer Martin Rees; and theoretical physicists Paul Davies and Lee Smolin.

Part One is an exploration of the future "in theory." Among the topics: advances in cosmology, the use in mathematics of "virtual unreality systems," new directions in complexity theory, speculations on what it means to be "alive," on how we learn, on how we think, on the nature of our consciousness and how we feel, on whether or not we are alone as a unique form of intelligence in this universe.

Part Two explores the future "in practice." It covers topics such as the future of DNA sequencing and what it will teach us about ourselves; the exploration of Mars and the search for extraterrestrial life; our command over matter; our intimate interaction with our machines and particularly our computers; the future outlines of cyberspace, neuroscience, and the way we raise our children; and the prospects for our continuing physical and mental well-being.

We are going through a rapidly accelerating epistemological sea change. We are using unprecedentedly powerful new tools, and in the process, as the Oxford biologist J. Z. Young pointed out in his BBC Reith Lecture in 1951, we are becoming those tools. What we have lacked until recently is an intellectual culture able to transform its own premises as fast as our technologies are transforming us.

The Next Fifty Years is part of this beginning, a place where empiricism and epistemology collide and everything becomes different—and where we begin to rethink our own natures and what kind of world we live in. That synergy exists in the work of the thinkers represented in this book and in their contributions to this book.

John Brockman
New York City
September 2001

THE FUTURE,
IN THEORY

LEE SMOLIN

❑

The Future of the Nature of the Universe

WE ARE ASKED TO PREDICT the state of our science fifty years from today. Fifty years is a long time, given the pace at which physics and cosmology have progressed over the last several hundred years. But perhaps it is not too long a time to make predictions that will not seem entirely stupid by then. If you look back over the history of science, you will see that often the important questions people were asking had been answered fifty years later. And yet the progress of science has usually been slow enough that people speak roughly the same language as their colleagues working in the same field fifty years earlier.

Let's look back fifty years, then, and note what the big questions were. My own list would include:

1) What is the nature of the strong force that holds atomic nuclei together?
2) What is the nature of the weak force responsible for radioactive decay?
3) Is the Steady State model of the universe right, or might there have been a Big Bang, as speculated by Gamow and other fringe figures?

4) Do protons and neutrons have any internal structure?

5) Why do the proton and neutron have slightly different masses, while the electron is much lighter than either? Why is the neutrino massless? What is the muon and who ordered it?

6) What is the relationship between general relativity and quantum theory?

7) What is the right way to understand the quantum theory?

I think we can confidently assert that now we know the answers to the first four questions. We are still working on the last three. But the first have not been forgotten; indeed, the methods by which those questions were answered form the basis of the training of a theoretical physicist today.

If we look back a hundred years, however, we find that we no longer care about many of the questions people were asking then. I'm not enough of a historian to write a list of questions asked by physicists at the turn of the last century, but they would likely have been more concerned with the properties of the ether than with the properties of atoms. There was no evidence for the existence of physical atoms until a few years later—and, indeed, in 1900 many physicists did not believe that atoms existed. Others, like Ernst Mach, thought the question was not a part of physics because atoms would never be observed. As for astronomy, there was no evidence in 1900 for the existence of galaxies apart from our own Milky Way, nor did anyone have any idea what made the stars shine. So while physicists of the early 1950s would probably have understood the questions that physicists are asking now, no one at the beginning of the twentieth century could have understood even the words that physicists were using in 1950 to talk to each other.

Sometimes science changes so little over fifty years that it makes sense to try to predict what we will know after that span. But there are periods when progress is faster and this is no longer the case. It seems that there is a horizon, somewhere between fifty and a hundred years into the future, beyond which it may be useless to speculate in any detail about the progress of science.

Let's take a moment to consider why this is so. It's probably partly because fifty years is about the length of a scientific career, from the beginning of studies until retirement. This, then, is the span of time over which the conservative tendencies built into the structure of scientific careers act to retard the progress of science. Science is hard, and we scientists prefer to have as good an understanding of what we're doing as possible; thus, unless forced to do otherwise, we prefer to work with techniques and ideas we already understand well. Another factor is that the careers of young scientists are often controlled by senior people nearing retirement, who are in many cases no longer active and therefore unfamiliar with new techniques. Career-savvy graduate students, no matter how imaginative, hesitate to work on something not understood by the powerful old men and women of their field. Thus, in order to think about what my science will be like in fifty years, I imagine what the brightest of my graduate students will be talking about at their retirement parties. My guess is that unless they are forced by data they cannot otherwise explain to make a revolution comparable to that of the early twentieth century, they will be using the language we've taught them. If that's the case, the present exercise may be useful—

though the romantics among us would rather anticipate a revolution than confirmation of our own beliefs.

One can also speculate on what was different about the sociology of science in the first half of the twentieth century to enable such enormous progress. Two credible answers come to mind: One is that it was possible for outsiders, such as Albert Einstein and Paul Ehrenfest, to publish in spite of not having university positions; another is that the generation that preceded the inventors of quantum theory was mostly wiped out in World War I, leaving the field open for Heisenberg, Dirac, and their friends.

This said, what will we know about fundamental physics and cosmology in fifty years? Rather than guessing, I propose a method that has a chance of reaching conclusions that won't look silly in the 2050s. I will list the most fundamental questions that are currently unanswered. Then I will ask what developments we may expect in experimental and observational science which will enable answers to them to be checked. I won't worry about theoretical developments, since there are already proposed answers to all of my questions and I assume that over a time span of fifty years we theorists will be able to adjust our theories, or invent new ones, in response to the data.

Here, then, is my list of the seven most important open questions in fundamental physics and cosmology:

1) Is quantum theory true as presently formulated, or will it need to be modified, either to have a sensible physical interpretation or to unify it with relativity and cosmology?

2) What is the quantum theory of gravity? What is the structure of space and time on the Planck scale (10^{-33}

cm, or twenty orders of magnitude smaller than an atomic nucleus)?

3) What explains the exact values of the parameters that determine the properties of the elementary particles, including their masses and the strengths of the forces by which they interact?

4) What explains the large ratios of scales we observe? Why is the gravitational force between two protons forty powers of ten smaller than their electrical repulsion? Why is the universe so big? Why is it at least sixty powers of ten bigger than the fundamental Planck scale? Why is the cosmological constant smaller than any other parameter in physics by roughly the same ratio?

5) What was the Big Bang? What determined the properties of the universe that emerged from it? Was the Big Bang the origin of the universe? If not, what happened before it?

6) What constitutes the dark matter and dark energy that make up between 80 and 95 percent of the density of the universe?

7) How did the galaxies form? What do the patterns we observe in the distribution of the galaxies tell us about the early evolution of the universe?

The first four of these questions continue, and deepen, unanswered questions from fifty years ago. The other three are new. Let us then ask whether the observations and experiments we'll be able to make in 2050 will be sufficient to test answers that theorists may propose to these questions. Of course, anything could be invented in fifty years. If my method is to be believable, we have to be conservative about the development of technology. I will thus consider only technology already existing or under development. In the latter case, I will consider only technology that a sizable fraction of experts believe will work in the next few years. However, for each technology existing or under development, I will assume that over the course of fifty years it will have been developed as far as possible, given only constraints imposed by the laws of physics or economics. Ordinary microscopes have a natural limit given by the wavelength of light, while telescopes have a natural limit imposed by the speed of light in a universe of finite age. Other technologies are more likely to be limited by financial considerations. We can safely assume that no experiment will be done whose cost (at that time) exceeds the defense budget of the United States. Let me hasten to add that I am not an expert in experimental physics or observational cosmology and I have not done a careful study of the limits involved. So my estimates will of necessity be very broad. Based on an extrapolation of current technology to its natural and financial limits, here is what I think we may hope for by mid-century.

We may begin with quantum theory. At present, powerful new techniques are being developed that promise to greatly extend the regime over which the quantum theory has been experimentally tested—techniques chiefly in aid of developing quantum computers. These are macroscopic devices that use quantum effects, such as superposition and entanglement, to do computations impossible for ordinary computers. A quantum computer requires those effects—which have so far been observed only for atomic systems—to work for macroscopic systems like the circuits of a computer, so the devices test those predictions of quantum theory that differ most strongly from classical theory.

Because it has been demonstrated that quantum computers could break all the codes now used by governments, militaries, and business, a lot of money is going into quantum computing. So it is safe to assume that as long as quantum mechanics remains true when extrapolated to macroscopic systems, there will be quantum computers in fifty years, and there are also likely to be quantum communication devices that make use of quantum states extending nonlocally worldwide. And if the present quantum theory is only an approximation of a deeper theory, experiments with quantum computers are likely to reveal this. It is thus reasonable to conclude that fifty years from now we will know the answer to the first question.

Let's turn next to cosmology. By mid-century we will certainly have a detailed picture of the history of the universe, based on observations using the full range of the electromagnetic spectrum, plus neutrinos, cosmic rays, and gravitational waves. The parameters in our current cosmological models will have been measured to high precision, and we will know lots of other facts about the universe, such as the number of black holes, the distributions in space and time of stars, galaxies, black holes, neutron stars, quasars, gamma ray bursters, and other objects. In fact, we will probably know more about the detailed history and properties of the universe than we know now about the history of the surface of our planet. At least in terms of familiarity with the whole range of its phenomena, we will truly be "at home in the universe."

The results will strongly constrain current theories of the early universe, such as inflation. We will also have a detailed picture of how galaxies and the patterns of galactic clusters and superclusters formed. Even without direct observations of dark matter, those observations will strongly constrain theories of the nature of dark matter and dark energy. By mid-century we may or may not have directly observed dark matter and dark energy and learned enough about them to confirm or refute the various theories that have been proposed about them.

Some readers will ask whether all these observations might confirm the Big Bang theory. To answer that question, we must distinguish between two meanings of Big Bang cosmology. I'll call the first the Expanding Universe cosmology: This is the theory that the universe has expanded from a much denser and hotter state for roughly 13 billion years. Among the key events in this story is the decoupling of light and matter when the universe had cooled enough for atoms to be stable. For roughly a million years before that, the universe was filled with a plasma, like the interior of a star. Since the transition, the universe has been filled with a very dilute gas, transparent to light, and all of the structures we see—stars and their planets, galaxies, galactic clusters—were formed. And almost all the chemical elements were formed since the transition, in stars; only helium and other light elements, like deuterium and lithium, were formed beforehand. I think it is unlikely that the outline of this story will be modified fifty years from now. We will know much more about the processes whereby stars, galaxies, and the elements formed, but all the evidence will still support the Expanding Universe theory.

It is also safe to say that we will have observations that strongly constrain our theories about what happened in the very early history of the universe. As we turn back the clock, the density and temperature increase. It is interesting to wonder how far back we can go to constrain theories by observation. By mid-century the part of the theory open to test is likely to extend back at least to the Planck time, a period so small that 10^{43} of them would fit into one second. Take, for example, the inflation hypothesis. Under a certain set of reasonable assumptions, predictions of this theory are testable in current observations of the fluctuations in the cosmic microwave background. These observations make up one of the greatest achievements of recent science. But even if current observations are compatible with inflation, there remain many open questions: The predictions of inflation are simple and may well be implied by other theories, so measurements that are more detailed may be needed to distinguish inflation from possible rival explanations of the present data. Moreover, there are many different versions of inflation, and further measurements are certainly needed to distinguish them. We hope to have those more detailed measurements of the cosmic microwave background not in fifty years but in five. So it is reasonable, if not certain, to predict that by mid-century it will be old hat to consider theories of the expanding universe testable by detailed observations all the way back from our present era to the Planck time.

But the Planck time is still not the origin of time. Very different from the Expanding Universe theory is the assertion that the Big Bang was the absolute beginning of the universe. Even if we know that the universe is hotter and denser all the way back to some point that may be a fraction of a second after some theoretical beginning, this does not prove that something didn't happen before that to set the expansion in motion. So it remains possible that the universe existed, possibly in some different form, a long time before the moment of the theoretical "Big Bang." To distinguish these kinds of hypotheses, let me refer to them as Origin of the Universe theories.

There are several Origin of the Universe theories now

under study, and they are all compatible with the Expanding Universe theory and therefore with all existing observations. Some of these, such as the Hartle-Hawking "wave-function of the universe," predict (or better, assume) that the Big Bang was really the origin of time. Others, such as the idea that new universes are created in the collapse of black holes, predict that there was a universe before the Big Bang, and that what happened then determined the properties of the universe that emerged from the Big Bang. It is possible, but by no means assured, that we will have evidence constraining these theories to the point that we will know whether there was something before the Big Bang or not. This evidence will have to come from using gravitational waves to probe the universe in the earliest period of the expansion. Nothing else can do this, because the universe at early times is opaque to all forms of radiation except gravitational waves. Gravitational wave astronomy is currently under development, but no gravitational waves have yet been observed. There are on the table proposals for space-based gravitational wave detectors that will in principle be able to use gravitational radiation to take a snapshot of the universe at the Planck time and thus distinguish different theories of the origin of the universe. It is possible but by no means certain that this technology will be in place by mid-century.

Let me turn now to elementary particle physics. Here the limitations are economic. If there are no breakthroughs in accelerator technology, then accelerators much bigger than those currently under construction with multibillion-dollar budgets will not be built. The problem is that the increase in energy with money is logarithmic, so to get roughly ten times more energy you need to increase the budget a hundredfold. It's therefore safe to predict that by mid-century, unless new technology has been invented, we will have probed at most two or three orders of magnitude higher in energy and smaller in scale than we have so far. But even this modest increase will lead to a host of discoveries, such as the Higgs boson, which is the particle believed to be implicated in setting the masses of all the observed elementary particles. We also ought to have confirmed or refuted the hypothesis of supersymmetry, which is a key element of string theory. All this is very important but will still leave us more than fifteen powers of ten from the Planck scale, which we must probe in order to directly test quantum theories of gravity. Does this mean that candidates for quantum gravity, such as string theory and loop quantum gravity, will still be untested fifty years from now?

Perhaps not! New technology has made it possible to probe the Planck scale. The reasoning is as follows: Some quantum theories of gravity predict that space and time have a discrete atomic structure. If this is indeed the case, it would modify the way in which a photon traveled through space—rather like the way light is dispersed or refracted when passing through water. This effect is extremely small, but it is cumulative: If a photon travels very long distances, the tiny effect is amplified. Luckily we can easily observe photons that have been traveling for billions of light-years, coming from highly energetic events like gamma ray bursts. In these and a few other cases, the effects predicted by certain theories of quantum gravity may be large enough to be observed. Indeed, there are already claims that certain effects already observed in the behavior of very-high-energy cosmic rays may be explained by the effect of the quantum spacetime structure of space at the Planck scale.

I should emphasize that here I am referring to present-day observations made with satellites designed to do other sorts of observations. With satellites designed for this purpose, the Planck scale could be probed in the next ten years. Improving these experiments requires mainly improvements in the precision of the detectors, and I know of no natural or economic constraints on the precision of detectors of high-energy photons which could prevent this new method from being used as a detailed probe of the structure of spacetime at the Planck scale. As for cosmic rays, there will be economic constraints: Cosmic rays become rarer as their energy increases, so it takes a bigger detector to probe further in energy. However, detectors already proposed and under development will be sufficient to study the range of energies relevant to distinguish the different quantum theories of gravity; and here, also, very good data will exist by 2050.

String theory, which is considered one of the leading candidates for a quantum theory of gravity, has been criticized for making few testable predictions. But it does make a few, and one of these bears on the results of such observations—that is, that the results must be consistent with space having a smooth structure. This is required by a symmetry of the theory called Lorentz invariance. Other quantum theories of gravity predict that this symmetry is broken or modified by small effects. The differences between the predictions of the different theories are well within the domain of what these new experiments could test. Thus it may happen that in a few years string theory, or one of its competitors, will have been ruled out by observations.

So here is the result of our quick survey of the experimental possibilities. It seems likely that we will have good data constraining the answers to the first two questions, concerning quantum theory and quantum gravity, and the last two questions, concerning cosmology and astrophysics.

We may or may not have data that allow us to test theories of the origin of the universe. Gravitational wave astronomy may reveal information coming from an era of the universe before the Big Bang, if there indeed was such a time, but given the uncertain state so far of gravitational wave astronomy, this is not assured.

There are then two remaining questions, numbers 3 and 4. These concern the reason that the parameters of particle physics, such as the masses of the elementary particles and the strengths of their interactions, take the values they do. Here the situation is unclear. We already have a lot of relevant data, but we cannot answer these questions at present. Developments in particle physics that will be testable by the next generation of particle accelerators may help us to understand why the elementary particles have the masses and interactions they do. But it is not at all certain that probing a few more orders of magnitude will give us enough insight to decide these questions.

Alternatively, it may be that these questions are answered by a quantum theory of gravity. This was the original motivation for string theory. But so far it doesn't look that way. Instead, the present theoretical evidence is that quantum theories of gravity like string theory are compatible with a wide range of choices for the properties of the elementary particles. This is because the answers we seek have to do with which solution of a particular theory describes our universe, rather than with the choice of which of the competing theories is correct. Each theory seems to have many solutions, and each one describes a possible universe.

This has led some of us to theories in which there are in fact many universes. Or rather, there is one universe but it contains many regions, each of which is like our own. Each begins in a big bang, after which it expands, taking on a structure dictated by the laws of physics. There are, roughly speaking, two kinds of such many-universe theories. The first just posits that the world consists of a large number of universes, in each of which the laws of nature, or at least the parameters of the laws, are chosen randomly. This generally goes under the name of the anthropic principle. The second kind of theory posits a process in which new universes are born as a result of the formation of black holes. This theory, called cosmological natural selection, works something like evolutionary biology, because the most common kinds of universes will be those that make the most copies of themselves.

Of these, the anthropic principle is not testable at all, while the natural-selection hypothesis is falsifiable but not provable—at least, with present technology. To prove rather than disprove this theory, we need a probe of the time before the Big Bang. Gravitational waves may do this for us, but, as I indicated, perhaps not within fifty years. So while the natural-selection idea may have survived all the tests thrown at it by 2050, it may still not have been proved.

Of course, new ideas and technologies may be invented which change the situation drastically. But we have agreed here to be conservative and consider only existing ideas and technologies. If I have to hazard a guess based on this conservative analysis, it would be that fifty years from now we will know the answers to at least five of the seven questions I listed. But it is anybody's guess whether we will know the answers to questions 3 and 4. That is, we may know the quantum theory of gravity, understand the nature of the Big Bang, have discovered and agreed on the correct formulation of quantum theory, and still not be able to answer a simple question that was posed as long ago as the

1930s: Why do the proton and neutron have almost the same mass, and why is the neutron the one that is slightly heavier?

❑

LEE SMOLIN, a theoretical physicist, is a founding member of and research physicist at the Perimeter Institute, in Waterloo, Ontario. He is the author of *The Life of the Cosmos* and *Three Roads to Quantum Gravity*.

MARTIN REES

❏

Cosmological Challenges: Are We Alone, and Where?

THE PRIME EXPLORATORY CHALLENGE of the next fifty years is neither in the physical sciences nor in (terrestrial) biology. It is surely to seek firm evidence for, or against, the existence of extraterrestrial intelligence. By 2050 it would be, I think, surprising if we did not understand how life began on Earth. We would then, even without direct extraterrestrial evidence, be able to assess how likely it was that life of some basic kind had emerged on other planets. But this leaves open a second question that could prove more intractable: "If simple life has emerged, what are the odds against its evolving into something that we would recognize as intelligent?"

An armada of space probes will be sent to Mars in the next decade to study its surface and eventually return samples to Earth. There are longer-term plans to search elsewhere in our solar system with robotic space probes—for instance, in the atmosphere of Saturn's giant moon Titan and in the ice-covered oceans of Europa, one of Jupiter's moons. If any of these instruments reveals that even the simplest form of life has originated independently at a second location within our own system of planets, this would imply that simple life must be widespread in the Galaxy and beyond.

Nobody now expects "advanced" life anywhere else in our own solar system, but our sun is just one star among billions in the Milky Way alone. Could planets orbiting other stars harbor life-forms far more interesting and exotic than anything we might find on Mars? Could they even be inhabited by beings we could recognize as intelligent? Even if primitive life is common throughout the universe, the emergence of advanced life might not be. At the moment, I think agnosticism is the only rational stance on this issue. We don't know enough about life's origins—still less about whether natural selection is "convergent" or would yield a quite different outcome if rerun on Earth—to say whether intelligent aliens are likely or unlikely.

Attempts to search for signals from intelligent extraterrestrials have had a hard time getting public money—even at the level of the tax revenues from a single science fiction movie—because the topic is encumbered by flaky associations with UFOs and so forth. But fortunately the pioneering efforts of the SETI Institute in California continue to expand, backed by hefty donations from visionary private benefactors.

I'm enthusiastic about these searches, and would remain so even if I believed there were heavy odds against success, because of the import of any manifestly artificial signal—even something as boring as a string of prime numbers. We would not know, of course, whether the transmission came from anything "conscious" or from an artifact left by a species that had become extinct long before the message was transmitted. But such contact would confirm that concepts of logic and physics had emerged elsewhere than in human brains. We would look at a distant star with renewed interest if we knew it was another sun, shining on a retinue of planets of which one had a biosphere as intricate and complex as our Earth's.

Such searches may of course be doomed to fail. This plainly would be in some ways disappointing. Failure need not, of course, imply that no other intelligences exist—they could be leading contemplative lives and doing nothing to reveal their presence. On the other hand, our cosmic self-esteem would be boosted if our tiny Earth were indeed a unique abode of intelligence: We could justly view it in a less humble cosmic perspective than it would merit if the Galaxy teemed with complex life. We could regard Earth as a "seed," from which life could disperse throughout space. There is plenty of time lying ahead. The entire Galaxy, extending for a hundred thousand light-years, could be greened in less time than it took for us to evolve from the first primates. When the sun dies, 5 billion more years will have elapsed; that is five times longer than it has taken for natural selection to lead from the first multicellular organisms to Earth's present biosphere, including us. During those eons of time, there could be even larger qualitative leaps. Future changes could indeed be much faster if artificially directed, so that they occur on a cultural or historical timescale. We can't predict what role life will eventually carve out for itself: It could become extinct, or it could achieve such dominance that it would influence the entire cosmos. The latter is the province of science fiction, but it can't be dismissed as absurd.

From Other Worlds to Other Universes?

The challenges of firming up some of these speculations in the next fifty years will seem less daunting if we look back at what has been achieved during the twentieth century. A hundred years ago, it was a mystery why the stars were shining. Moreover, we had no concept of anything beyond our Milky Way, which was assumed to be a static system. In the last three decades, space probes have beamed back pictures from all the planets of our solar system and giant telescopes have allowed astronomers to look far deeper into space than was previously possible. Our cosmic panorama extends millions of times beyond the remotest stars we can see—out to galaxies so far away that their light has taken 10 billion years to reach us. Cosmic history can be traced back to within a second of the beginning. Only in the first millisecond of cosmic expansion and deep inside black holes do we confront conditions where the basic physics remains unknown.

These advances have raised a fascinating possibility: that what we call our universe—the domain that astronomers can probe and which originated in the Big Bang—may not be all of reality. Theorists have already come up with illustrative scenarios for multiple universes, based on assumptions that are well defined but speculative. Andrei Linde, Alex Vilenkin, and others have performed computer simulations depicting an "eternal" inflationary phase in which many universes sprout from separate big bangs into disjoint regions of spacetime. Alan Guth and Lee Smolin have, from different viewpoints, suggested that a new universe could sprout inside a black hole, expanding into a new domain of space and time inaccessible to us. And Lisa Randall and Raman Sundrum conjecture that other universes could exist separated from us in an extraspatial dimension; these disjoint universes may interact gravitationally, or they may have no effect whatsoever on one another. In the hackneyed analogy wherein the surface of a balloon represents a two-dimensional universe embedded in our three-dimensional space, these other universes would be represented by the surfaces of other balloons. Any bugs crawling around on one such surface, and with no conception of a third dimension, would be unaware of their counterparts crawling around on the surfaces of other such balloons. Other universes would be separate domains of space and time. We couldn't even meaningfully say whether they existed before, after, or alongside our own, because such concepts make sense only insofar as we can impose a single measure of time ticking away in all the universes.

Alan Guth and Edward Harrison have even conjectured that universes could be made in the laboratory, by imploding a lump of material to make a small black hole. Is our entire universe perhaps the outcome of some experiment in another universe? Smolin speculates that a daughter universe may be governed by laws that bear the imprint of those prevailing in its parent universe. If so, the theological arguments from design could be resuscitated in a novel guise, further erasing the spurious boundary between natural and "supernatural" phenomena.

Parallel universes are also invoked as a solution to some of the paradoxes of quantum mechanics, in the "many worlds" theory, first proposed by Hugh Everett and John Wheeler in the 1950s. This concept was prefigured by the visionary science fiction pioneer Olaf Stapledon, as one of the more sophisticated creations of his Star Maker:

Whenever a creature was faced with several possible courses of action, it took them all, thereby creating many . . . distinct histories of the cosmos. Since in every evolutionary sequence of the cosmos there were many creatures and each was constantly faced with many possible courses, and the combinations of all their courses

were innumerable, an infinity of distinct universes exfo-
liated from every moment of every temporal sequence.

None of these scenarios has been dreamed up out of the
air; each has a serious, albeit speculative, theoretical motivation. However, only one of them, at most, can be correct.
Quite possibly none is; there are alternative theories that
would lead just to one universe.

Firming up any of these ideas will require a theory that
consistently describes the extreme physics of the very beginning, when our current universe was squeezed so small that
quantum fluctuations could shake it—a theory, in other
words, that reconciles Einstein's theory of gravity (general
relativity) with the quantum principle, which governs the
microworld of subatomic particles.

Einstein himself spent his last thirty years seeking such a
unified theory, in vain. Unkind critics said that he "might
as well have gone fishing" for the last half of his life; now
we can see that his quest was premature, because he was
unaware of the nuclear and weak forces which any such
theory must incorporate along with electromagnetism. Unified theories are likely to offer a more realistic challenge in
the twenty-first century than they ever were in the twentieth, and the time now seems ripe for a realistic assault on
them. But consistency is not enough; there must be grounds
for confidence that a unified theory isn't a mere mathematical construct but applies to external reality. We would
develop such confidence if the theory accounted for things
we can observe that are otherwise unexplained—such as a
link between the nuclear, electric, and gravitational forces,
or why there are three kinds of neutrinos. In the coming
decades, physicists may well develop such a theory. If so, it
will be the end of an intellectual quest that began before

Newton and continued through Maxwell, Einstein, and their successors. But I'd be even more excited if it offered insight into why our universe has features which, to some of us, seem surprising.

Our Special, Biophilic Universe

If aliens exist in our universe, they may understand and "package" aspects of reality very differently from the way our own minds do, but if we ever establish contact we're assured one common interest. They will be made of similar atoms and governed by the same physical laws. If they have eyes and clear skies, they will gaze out on the same vista of stars and galaxies as we do. We all trace our origins back to a common genesis—a Big Bang about 13 billion years ago.

But our existence—and that of the aliens, if there are any—depends on our universe's being rather special. A universe hospitable to life—what we might call a biophilic universe—must, it seems, be adjusted in a particular way. The prerequisites for any form of life—long-lived stable stars, atoms (such as carbon, oxygen, and silicon) able to combine into complex molecules, and so on—are sensitive to the physical laws and to the size, expansion rate, and contents of the universe. If the recipe imprinted at the time of the Big Bang had been even slightly different, we could not exist. Many recipes would lead to stillborn universes, with no atoms, no chemistry, and no planets—or to universes too short-lived or too empty to allow anything to evolve beyond sterile uniformity. This distinctive recipe seems a fundamental mystery.

A profound question that exercised Einstein was "Did God have any choice in the creation of the world?" He was pondering—in poetic language—whether the laws governing our universe are unique, for some deep mathematical reason, or whether the recipe could have been fundamentally different. If our universe was the unique outcome of a fundamental theory, we would have to accept the biophilic tuning as a brute fact. On the other hand, if the answer to Einstein's question is yes, then the underlying laws could be more permissive: They may allow many recipes, leading to many different kinds of universes. The entire multiverse would be governed by a set of fundamental principles, but what we call the laws of nature (or, at least, some of them) would be no more than local ordinances, the outcome of environmental accidents in the initial instant after our own particular Big Bang.

As an analogy, consider a set of snowflakes. They all have one thing in common: hexagonal symmetry. This symmetry is a consequence of the shape of the water molecules that the flake is made of. But it is hard to find two that look the same. Each flake has a pattern that depends on its distinctive history—how, in detail, the ambient temperature and pressure changed as it fell through the cloud where it grew. Some features of our universe, likewise, may be accidental consequences of how it cooled down after the Big Bang— rather as a piece of red-hot iron becomes magnetized when it cools down, but with an alignment that may depend on chance factors—instead of having been imprinted at a deeper and more fundamental level that obtains throughout the entire multiverse. If physicists achieve a convincing fundamental theory, it should tell us which aspects of nature are direct consequences of the bedrock theory (just as the symmetrical template of snowflakes is due to the basic structure of a water molecule) and which are (like the distinctive pattern of a particular snowflake) the outcome of accidents.

If there is indeed an ensemble of universes (and a theory capable of describing the very beginning of our Big Bang might settle that issue as well), then most universes would be sterile—governed by laws that preclude complex structures—or too small or short-lived to allow space and time for evolution to do its work in generating complexity. We (and any aliens that may exist) would find ourselves in one of the small and atypical subsets governed by laws that permit complex evolution. The seemingly "designed" features of our universe shouldn't surprise us, any more than we are surprised at our particular location within our universe. We shouldn't take the Copernican "principle of mediocrity" too far. We find ourselves on a planet with an atmosphere, orbiting at a particular distance from its parent star, even though this is a very special and atypical place. A randomly chosen location in space would be far from any star—indeed, it would most likely be somewhere in an intergalactic void, millions of light-years from the nearest galaxy. If there are many universes, most of which are not habitable, we should not be surprised to find ourselves in one of the habitable ones.

We may one day have a convincing theory that tells us whether or not a multiverse exists and some of the so-called laws of nature are just parochial bylaws in our cosmic patch. Even before we have such a theory, we can test the plausibility of "anthropic selection" by asking if our actual universe is typical of the subset in which we could have emerged. If it is a grossly atypical member even of this subset, and not merely of the entire multiverse, then we would need to abandon the multiverse hypothesis. As another example of how multiverse theories can be tested, consider Smolin's conjecture that new universes are spawned within black holes and that the physical laws in the daughter universe retain a memory of the laws in the parent universe, as if there were a kind of cosmic heredity. If Smolin is right, universes that produce many black holes would have a reproductive advantage, which would be passed on to the next generation. Our universe, if an outcome of this process, should therefore be near-optimum in its propensity to make black holes, in the sense that any slight tweaking of the laws and constants would render black-hole formation less likely. Smolin's concept is not yet bolstered by any detailed theory of how any physical information (or even an arrow of time) could be transmitted from one universe to another. I personally hold little hope for its survival. However, Smolin deserves our thanks for providing a demonstration that a multiverse theory can in principle be vulnerable to disproof.

These examples show that some claims about other universes may be refutable, as any good hypothesis in science should be. We cannot confidently assert that there were many big bangs—we just don't know enough about the ultra-early phases of our own universe. Nor do we know whether the underlying laws are permissive; settling this issue is a challenge for physicists in the next fifty years. But if the underlying laws are permissive—if they allow for the creation of all sorts of universes—then so-called anthropic explanations for why ours is the way it is would become legitimate; indeed, they would be the only type of explanation we will ever have for some important features of our universe. Parts of cosmology would become like evolutionary biology.

What we have traditionally called the universe may be the outcome of one big bang among many, rather as our solar system is merely one of many planetary systems in the Galaxy. Just as the pattern of ice crystals on a freezing pond is an accident of history rather than the result of a fundamental property of water, so some of the seeming constants of nature may be arbitrary details rather than being uniquely defined by the underlying theory. The quest for exact formulas for what we normally call the constants of nature may consequently be as vain and misguided as was Kepler's quest for the exact numerology of planetary orbits. And other universes will become part of scientific discourse, just as "other worlds" have been for centuries. Nonetheless (and here scientists should gladly concede ground to philosophers), any understanding of why any-thing exists—why there is a universe (or multiverse) rather than nothing—remains in the realm of metaphysics and undoubtedly always will.

❑

SIR MARTIN REES is Royal Society Professor at Kings College, Cambridge. He was previously Plumian Professor of Astronomy and Experimental Philosophy at Cambridge, having been elected to this chair at the age of thirty, succeeding Fred Hoyle. He has originated many key cosmological ideas: For example, he was the first to suggest that the fantastically energetic cores of quasars may be powered by giant black holes. For the last twenty years, he has directed a wide-ranging research program at Cambridge's Institute of Astronomy. He is the author of several books, including *Gravity's Fatal Attraction* (with Mitchell Begelman); *New Perspectives in Astrophysical Astronomy; Before the Beginning: Our Universe and Others; Just Six Numbers: The Deep Forces That Shape the Universe;* and, most recently, *Our Cosmic Habitat.*

IAN STEWART

❑

The Mathematics of 2050

AMONG ALL THE SCIENCES, mathematics probably has the longest unbroken history—rivaled only by astronomy. Both go back at least to the ancient Babylonians, and discoveries made then remain important today. And just as astronomy builds on the discoveries of the past, so does mathematics. Astronomy is based on observations of the real world, whereas mathematics is a shared social construct of ideas; but ideas are the driving force of astronomy, and mathematics has grown out of attempts to model the real world—to count the passing days, to measure the sizes of fields, to calculate the taxes owed to the king.

In astronomy, there have been revolutions. Old ideas have been overthrown, replaced by radically different new ones. In 1877, for example, the Italian astronomer Giovanni Schiaparelli saw *canali* ("channels") on Mars; the mistranslation of this observation quickly led to a widespread belief (even among astronomers) that Mars was inhabited by intelligent beings. Now we know better. It is often said that mathematics cannot have revolutions, because the nature of mathematical truth does not change. But human attitudes do change, and one of the biggest revolutions in mathematics was a major revision of our concept of mathematical "truth." Thanks to Kurt Gödel and Alan Turing, we now realize that even mathematical truth is not absolute.

The next fifty years will witness several major revolutions in mathematics. Some are already under way—the increasing influence of the computer and the new challenges posed by the biosciences and the financial sector. There will be others, but the only safe prediction is that many will be unpredictable. Various commentators have predicted a change in the very notion of proof, which is the central concept of mathematics. Some of them say that the computer will bring about a fundamental revision of the concept of proof, others that the concept will die out altogether. Both of these views are based on fundamental misunderstandings of current trends. In mathematics, proofs are the substitute for the observations and experiments of the rest of science—that is, the means whereby its practitioners avoid getting sidetracked by their own cleverness and believing things to be true because humans would like them to be true. The computer no more dispenses with the need for proof than the invention of the microscope dispensed with the need for experiment in biology. As in this biological analogy, the computer has modified and enhanced the techniques of proof, but it has not altered the underlying philosophy—that a proof is a logically coherent story, deriving new theorems from existing ones in a manner that can withstand the closest scrutiny by skeptical experts. The concept of proof and the belief that it is essential to the mathematical enterprise will survive the next fifty years unscathed.

Mathematics derives its power from the combination of two different sources. One is the "real world." Kepler,

Galileo, Newton, and others have taught us that many aspects of the external universe can be understood as the working out of simple but subtle mathematical rules, "laws of nature." Occasionally physicists revise their formulations of these laws. Newtonian mechanics gives way to quantum mechanics and general relativity; quantum mechanics gives way to quantum field theory; quantum gravity and super-strings point in the direction of a future revision. Problems in the real world stimulate the invention of new mathematics, and the mathematics usually survives and continues to be important even when the theory that gave birth to it is changed.

The second source of mathematics is the human imagination—the pursuit of mathematics for its own sake. The path from the real world to a fully developed branch of mathematics is rocky: A certain amount of exploration is useful; every so often intrepid pioneers depart from the current path in search of their own private visions and return having discovered a far better route. To the pioneers, the value of their explorations is obvious: It is what drives them, and it needs no justification beyond its own intrinsic interest.

These two styles of mathematics are usually character-ized as applied mathematics and pure mathematics. Neither phrase is accurate, and both are open to misconceptions. Many areas of "applied" mathematics are not actually applied to anything real; the purity of "pure" mathematics refers to its methods, not to a disdain for the discipline's practical value. But the terms do mark the extremes of a spectrum of mathematical styles—a continuum that con-nects the imperatives of the outside world to the enigma of the human imagination. It is this continuum and the two-way traffic of ideas along it that give mathematics its power. We need both styles if we are to make progress, and it is pointless to assert the superiority of one or the other.

A hundred years ago, most mathematicians spanned the continuum. Fifty years later, the continuum had become too big for any one mind to grasp, and individuals tended to specialize, leading to the discipline's apparent fragmentation. The pure mathematicians and the applied mathematicians separated into two camps, each with a different philosophy. They disagreed about fundamentals, about the need for proof, about methods, about what problems were interesting. They were two sundered sects of what had once been a broad, inclusive church. But by the turn of the millennium this self-destructive tendency had reversed. Methods from pure mathematics have infused new life into applied mathematics; problems arising in applications have stimulated new developments in pure mathematics. The dividing line, always owing more to ideology than reality, has begun to blur. Over the next fifty years, the trend toward greater unification will accelerate, and soon there will be just mathematics, with no qualifying adjectives and no sectarian disputes. There will still be specialists, but their specialties will combine the abstract logic and conceptual emphasis of pure mathematics with the concrete concerns of applied mathematics. We will all be mathematicians, all part of the same great endeavor, dabbling in our own little patches of the great collective "extelligence" of mathematical thought. We will be aware of the existence of other dabblers in other patches, grateful for their existence, and respectful of their activities as valid contributions to the enlargement of the whole.

One prediction that can confidently be made about the next fifty years is that we will see huge advances. The golden age of mathematics is not ancient Greece or Renaissance Italy or Newtonian England, but now. And in fifty years' time it will still be now. A good test of this contention is progress on big unsolved problems, questions asked hundreds of years ago that continued to baffle the greatest minds until one day an entry route was found, a new idea was brought to bear, and the problem cracked wide open. The best known example of recent times is Andrew Wiles's proof of Fermat's last theorem. Sometime around 1637, Pierre de Fermat wrote, in the margin of his personal copy of the *Arithmetica* of Diophantus, that two perfect cubes cannot add to form a perfect cube, and similarly for fourth powers or higher. The problem of proof resisted all efforts until 1995, when Wiles pulled off one of the greatest mathematical coups of the twentieth century. His solution involved a new method: to translate Fermat's statement into a far broader one about "elliptic curves"—a very different area of number theory—and then to bring every possible modern weapon to bear on the resulting question.

The most celebrated unsolved problem at this moment is the Riemann hypothesis, first stated by Georg Bernhard Riemann. This is a rather technical question in complex analysis whose conjectured answer would shed a great deal of light on prime numbers, algebraic number theory, algebraic geometry, even dynamics. In recent years, intriguing connections have emerged with quantum physics. I will stick my neck out and predict that by 2050 the Riemann hypothesis will have been proved—the expected answer will turn out to be right—and that links to physics will play a major role in the proof. But I'll hedge my bets by predicting that the final route to a solution will be based not on the current links to physics but on a connection not yet envisaged.

In 1900 David Hilbert, the leading mathematician of his

age, laid down twenty-three significant problems for the future to solve. Most have been disposed of, but not the Riemann hypothesis. In 2000 the Clay Mathematics Institute, in Cambridge, Massachusetts, offered prizes of $1 million each for solutions to seven long-standing and intractable mathematical problems. One is the Riemann hypothesis. The others are the Poincaré conjecture, a topological characterization of the three-dimensional sphere; the P/NP problem of theoretical computer science, which asks for a proof that difficult computations really exist; the Hodge conjecture and the Birch/Swinnerton-Dyer conjecture in algebraic geometry; the existence (or not) of solutions to the Navier-Stokes equations of viscous fluid dynamics; and a proof of the "mass gap hypothesis" in quantum field theory. I suspect that by 2050 we will know a lot more about all seven problems, with mixed results. At a guess, the Poincaré conjecture will still be wide open, the P/NP problem will be proved formally undecidable, the Hodge conjecture will be disproved, the Birch/Swinnerton-Dyer conjecture will be proved, the Navier-Stokes equations will turn out not to have solutions in certain exotic circumstances, and the mass gap hypothesis will have been settled one way or another but physicists will no longer be interested in it.

The existence of $7 million in prize money will not have diverted mathematics onto new paths. That wouldn't work in any case, since mathematicians are not especially attracted by cash rewards, unlike molecular biologists. But it will have achieved its aim of signaling to the outside world how important these seven questions are—and, by extension, mathematics in general. I would like to predict that this message will have got through to government funding agencies, who will finally have realized that a billion dollars spent on mathematics would transform human existence far more substantially, and to more positive effect, than the same sum spent on a few bits and pieces for a new particle accelerator or yet another massive exercise in biological stamp-collecting. I'd like to, but I won't.

The P/NP problem is about computers, but it will not be solved by a computer. What it needs is a good old-fashioned idea. Computers won't help with that particular problem even in an exploratory way; but their use will play another role, an increasingly indispensable one, in informing mathematicians of possible conjectures, which they will then seek to prove. More than that, computers will also play a central role in many proofs, a trend that is already under way. With appropriate programming, computers can be much more than the number crunchers of the 1960s. Already they are being used in a logically rigorous manner to "assist" in proving theorems. The best-known example is the 1976 proof by Kenneth Appel and Wolfgang Haken of the four-color theorem. This theorem, first stated in 1852 by Francis Guthrie, says that any map in the plane can be colored using four colors so that no two adjacent regions have the same color. The conceptual part of the proof was to reduce the theorem to a routine verification that some two thousand special maps, which were found without use of a computer, possessed a particular mathematical property. The computer then did the necessary calculations to check that this was so.

Some philosophers argue that computer-assisted proofs are qualitatively different from traditional ones, because the calculations cannot be checked by an unaided human being. However, we can ask why the human checking of proofs was introduced in the first place: The point is that what matters is the checking, not what kind of entity carries it out. In the past, humans did the checking, because that was the only alternative, but it doesn't have to be humans in the future. The main criteria are that the checking entity must be trustworthy and anyone who distrusts it should have recourse to independent entities capable of carrying out their own checks. But as long as those conditions apply, the decision of a machine is just as valid as that of a human being. Most mathematicians expect a computer to make fewer errors of computation or of logic than a human when it comes to routine, boring calculations. Indeed, the history of the four-color theorem is littered with errors made by humans. What counts is the logic of the computer program, and whether the machine is really behaving as its designers intended. Both of these matters can be checked independently. The "thinking" part of the work is still being done by humans—recasting the problem in a form that reduces it to a huge but routine list of calculations. After that, using a computer is philosophically no different from using a book of mathematical tables as a calculating aid.

This tendency to use computers as exploratory aids and in proofs—analogous to the biologists' use of the scanning tunneling microscope and gene-sequencing equipment—will be ingrained in the mathematics of 2050. There will be "virtual unreality" systems, allowing mathematicians to "visit" abstract conceptual structures such as non-euclidean geometries or ranges of giant primes and manipulate them at will, almost without effort—just as we now use calculators to do arithmetic. Already the ingredients for VU are being assembled. Soon they will come together. The needs of software engineers will stimulate new developments in combinatorics—"finite mathematics." Today's uneasy and occasional interactions between combinatorics and geometry will gel into an intimate relationship, brought into being by the relation between circuit layout and logical function.

In Newton's day, the main external sources of mathematical problems were astronomy and mechanics, the physical sciences. By 2050, more exotic sciences will be tied into mathematics in the same way. One will be quantum physics, already highly mathematical. Today, surprising new links between quantum field theories, geometry, topology, and algebra are beginning to be uncovered. Many more will follow. Within fifty years, new structures inspired by quantum field theory, superstrings, and whatever lies beyond will have created entirely new areas of algebra and topology. The mathematicians of the nineteenth century generalized traditional "real" numbers into "complex" numbers, where −1 has a square root, and this turned out to be an extraordinarily prolific idea. Very quickly, every area of mathematics was "complexified": A fruitful complex analog of the old real-number mathematics was devised. Quantization will be the twenty-first-century version of complexification; we will be doing quantum algebra, quantum topology, quantum number theory.

Far more influential, and far more radical, will be the mathematics inspired by the biosciences: biomathematics. As the triumphal announcements about the human genome give way to a new realism about the results, it has become clear that merely sequencing DNA does not get us very far in understanding organisms, or even in curing diseases. There are huge gaps in our understanding of the link between genes and organisms. And the sequenced genome tells us virtually nothing about how to manage our ecosystems, like coral reefs and rain forests. Confident predictions that the human genome would have a hundred thousand genes have been confounded: there are a mere thirty-four thousand. The "map" from genes to proteins is far more complicated than expected; indeed, it probably isn't a map at all. Genes are part of a dynamic control process that not only makes proteins but modifies them and gets them to the right place in a developing organism at the right moment in its life history. The understanding of this process will require much more than a mere list of DNA codes, and most of what's missing has to be mathematical. But it will be a new kind of mathematics, one that blends the dynamics of organism growth with the molecular information-processing of DNA. The DNA code remains important; it just isn't everything. The new biomathematics will be a strange new mixture of combinatorics, analysis, geometry, and informatics. Plus lots of biology, of course.

A straw in the wind in this direction is the rapid growth of the science of complex systems: that is, systems formed by large numbers of relatively simple components—"agents" or "entities"—that interact in simple ways. We have learned that the apparent simplicity is deceptive: Out of it come high-level patterns, "emergent phenomena." Out of the connectivity of human brain cells comes conscious awareness, for example. By 2050 we will have a rigorous mathematical theory of emergent phenomena and the high-level dynamics of complex systems. It will lead to concepts not yet dreamed of, but also to a new understanding of the limitations of mathematical modeling in science. Today, complex systems are being studied in two main areas—biology and finance. A stock market, for instance, has many agents who interact by buying and selling stocks and shares. Out of this interaction emerges the financial world. The mathematics of finance and commerce will be revolutionized by throwing away the current "linear" models and introducing ones whose mathematical structure more accurately reflects the real world.

Even more dramatically, mathematics will invade new areas of human activity altogether—social science, the arts, even politics. However, mathematics will not be used in the same way as it is currently used in the physical sciences. In physics, mathematics is used to state quantitative laws, and predictions about the real world are often the result of gigantic computations, in which the link from the laws to the resulting patterns becomes impossible for the human mind to follow. For example, the vast spiral of a hurricane is modeled by writing down equations for the motion of billions of tiny regions of warm moist air and then performing huge calculations to solve those equations. An alternative approach, currently in its infancy, would be to deduce the spiral shape from the general structure of the equations—their symmetries, for example. A kind of "calculus of spirals" would replace the interminable manipulation of numbers. More generally, we can hope to see the beginnings of a qualitative, contextual theory of dynamic pattern formation.

Finally, mathematics will help us to understand the patterns of the universe in terms of the patterns themselves and not just in terms of billions of dancing digits out of which the patterns emerge like some kind of miracle.

❑

IAN STEWART is the 1995 recipient of the Royal Society's Michael Faraday medal for outstanding contributions to the public understanding of science. He has written numerous articles on mathematics for such popular magazines as *Discover, New Scientist,* and *The Sciences.* For ten years he wrote the "Mathematical Recreations" column in *Scientific American*, and he is mathematics consultant to *New Scientist*. He is also coauthor (with Jack Cohen) of *The Collapse of Chaos* and *Figments of Reality* and author of *Does God Play Dice?*, *Fearful Symmetry*, *From Here to Infinity*, *Nature's Numbers*, *Life's Other Secret,* and *Flatterland.*

BRIAN GOODWIN

❑

In the Shadow of Culture

LOOKING INTO THE FUTURE at this point in history is as
difficult as it must have been in 1600. Then the Western
world's feudal system had virtually disintegrated, except
for its monarchies. The imperial order of the Holy Roman
Empire was dissolving under the dual impact of emerging nation-states and Protestant sects, and in the coming
decades the Thirty Years' War would plunge Europe into a
new Dark Age. Shakespeare was still writing his plays cele-
brating the remarkable complexity and diversity of human
nature, his characters emerging from a world described by
Renaissance magi in which celestial harmony was experienced through the music of the spheres and love made
the world go round. Galileo was rolling his cylinders down
inclined planes and struggling to understand the erratic
motions of the moons of Jupiter; soon the Church would
discipline him for claiming in support of Copernicus that
the earth really does move around the sun rather than simply appear to do so from a mathematical perspective, and
he would reluctantly recant.

Francis Bacon extolled the virtues of Galileo's method

of understanding nature, but the scientific approach to knowledge was in shadow. The church's teaching about the hierarchy of beings, together with the magical worldview found in Shakespeare, constituted the dominant understanding of the order of things and humanity's place in the cosmos. Who could have predicted that within fifty years what lived in the shadows in 1600 would have become the basis for the new cultural orientation that we have come to know as modernity? Galileo's method of observation, measurement, and mathematical relationships offered reliable knowledge of nature to a culture from which all certainties had been stripped by the fragmentation of other knowledge systems and the devastation of the Thirty Years' War. Isaac Newton, through his theories of planetary motion in which gravity displaces love as the force that makes the world go round, would soon convince all doubters that science was the path to understanding.

We have now reached another cultural terminus. Given the unexpectedness of the seventeenth-century transition to the modern age, from which we are now emerging into another cultural epoch, it seems futile to suggest what lies in store fifty years into the future. However, there is a way to prepare for the unexpected so that the appropriate transition is facilitated even if it cannot be foreseen. This involves doing the opposite of what this collection of essays proposes. I suggest that we take our focus off the next fifty years in order to examine the present and experience it as fully as possible, particularly those aspects that are in shadow and just beginning to come into the light. It may then be possible to feel our way into a creatively emergent future, even if we don't know where we're going.

The Visible

What stands out clearly today is the powerful alliance of science, technology, and business which has created a global culture whose primary principles are based on prediction, control, innovation, management, and expansion. Underlying these principles are rationality and power, which Francis Bacon championed as the path to understanding nature and using that knowledge to liberate humanity from bondage. This strategy has worked extraordinarily well. We do indeed have the means to liberate all human beings from hunger and poverty, with the production of wealth and goods achieved through the application of scientific knowledge and the expansion engine that drives capitalism. However, things have not worked out quite as predicted. A large and increasing fraction of the world's population still lives in hunger and poverty; agricultural land and natural resources are being destroyed at an increasing rate; the pollution of land, sea, and air is affecting all life on the planet; the atmosphere grows more and more turbulent as a result of global warming caused by the burning of fossil fuels; species are going extinct at a rate not seen since the extinction spasms of the Permian and the end of the Cretaceous; and the ability of nation-states to protect their citizens dwindles with the rise of transnational organizations that enforce unregulated global trade in goods and services. The extraordinary expansion of our information technology has created conditions wherein decisions about investment and movement of capital can destabilize markets and even dissolve governments. The increasing chaos of the weather is reflected in widespread political chaos, in which traditional discourse seems powerless to achieve stability and security—which is what science and technology, the gifts of modernity, had offered us in the first place. Unexpectedly, we have been plunged into a Dark Age much more dangerous than the Thirty Years' War, because the disintegration is now global.

All these signs are highly visible. We can use such an image of the present to suggest either impending apocalypse or a transformation to a brighter future based on new coherences and alliances of present trends. My intention is to do neither, but instead to try to identify aspects of our present situation that are less visible but seem to be emerging. My goal is not to describe the next fifty years but to see what is dimly present and might prompt creative action now, without a clearly defined vision of the future.

The Invisible

Since I am a scientist, and science is likely to contribute significantly to our future, I am going to reflect largely on what lurks in the wings of the scientific establishment as a possible player in what is to come. My first story resonates with Galileo's experience at the hands of the church, which made him recant his support of the heresy that Earth moves around the sun.

In the 1960s the scientist/inventor James Lovelock, who was working with NASA on issues relating to extraterrestrial life, had the insight that the composition of Earth's atmosphere distinguishes it from the other planets in a manner that tells us something profound about the relationship between living organisms and their inorganic environment. He wrote an article in the science journal *Nature* arguing that life doesn't simply adapt to given conditions on the planet where it takes root; it changes those conditions and stabilizes them so as to perpetuate itself. This insight, supported and extended by biological evidence from the work of Lynn Margulis on the power of microbes to alter planetary conditions, was presented to the scientific world as the Gaia hypothesis in a 1974 article coauthored by Lovelock and Margulis in the journal *Tellus*. Here was science based on sound evidence but dressed in the garb of the ancient Greek earth goddess. What did the scientific community make of it? They cast the hypothesis into the outer darkness. Why? Because Lovelock and Margulis had violated not one but two principles of orthodox science. The first violation was the suggestion that there are basic aspects of evolution that do not conform to Darwinian principles; according to the Gaia hypothesis, life does not simply adapt to given conditions on Earth but can change those conditions so that they are adaptive to life. For instance, microbes can alter the composition of the atmosphere (CO_2, CH_4, NH_3, O_2) so that the temperature remains within bounds that allow life to continue. The whole earth can thus be seen as a living system, regulating its own vital variables as an organism does.

The second violation was the use of the term "Gaia" in the hypothesis, which implied that Earth itself is a kind of living being rather than a set of blind, mechanical processes of the type that science sees as carrying out planetary activity. This image of Earth as goddess was an especially powerful one for environmental activists, who were protesting the pillage of Earth's natural resources and the pollution of land, sea, and air by (for example) excessive burning of fossil fuels. The Gaia image provided a focus for the feelings of grief and anger that many people were experiencing when they realized what we had done and were still doing to our planet.

Gaia's chief proponent, Lovelock, was effectively excommunicated from the church of science as a result of the heresies in the hypothesis. He responded with a vigorous defense of his first heresy, agreeing that Darwin's principle of natural selection was one of the mechanisms of evolution but insisting that life itself changed conditions on Earth as well as adapting to change. This principle has now been accepted by the scientific community, which uses the term "Earth system science" to describe the extended picture of terrestrial evolution that Lovelock and Margulis introduced. But acceptance came at a price: Lovelock recanted with respect to the implication that Earth has any qualities of intent or purpose of caring for life on the planet—a notion that was seen as a kind of animism, which is absolutely forbidden in science. If it lives anywhere in our culture, it is in the deepest shadow.

Dead or Alive?

Animism, the view that everything in creation is alive in some sense, is not only ostracized from science but is outside the general belief system of our culture. I talked recently with an accomplished flautist in the Santa Fe Opera orchestra about what I had learned from a Navajo flautist about the seven flutes he carries with him when he performs. He had demonstrated the different qualities of tone and expression of each, and when I pointed to one whose birdlike quality I found particularly beautiful, he played it again and then gently explained that in his culture it was considered rude to point at a flute, just as we learn that it is rude to point at a person. It was rude because each flute had a name and a personality; a flute was actually a living being. The Opera orchestra flautist looked at me in disbelief. "Was he serious?" she asked. Much as she loved her own flute, it was in no sense alive for her.

Why is animism so threatening to the Western scientific worldview? Is there any sign that the dialectic of science is beginning to bring this view into the light again? And if so, what might be the future significance of this resurrection? Our science insists that the energy/matter that is the stuff of the cosmos is dead, without any kind of sentience. Galileo learned this lesson from the Greek atomists: There is nothing but atoms and the void. This way of looking at things has allowed us to explore many aspects of the world and develop the most extraordinary insights into the nature of the processes underlying the diverse phenomena of nature, both dead and alive. And it has given us an impressive range of technologies. This way of knowing works; it is reliable. The basic proposition is that those aspects of nature that can be quantified, measured, and preferably organized into mathematical relationships that describe their regularities of behavior provide the only certain and objective knowledge of the world. When we describe phenomena not in quantitative terms but in terms of their qualities, such as the joyful playfulness of an otter, the beauty of a landscape, or the vitality of a friend, we are expressing our own personal view of what we are observing. Playfulness, joy, beauty, and vitality are all qualities without quantifiable measure, so they cannot be used as a basis for reliable descriptions of phenomena. They may well have quantifiable aspects that could be used for scientific description, but the qualities themselves lie outside science, in the realm of subjective experience. To say that a flute is alive and has feelings that can be violated by pointing at it makes no sense (other than metaphorically) from this perspective. A flute is not only dead, it is void of experience of any kind.

Where Does Consciousness Come From?

One of the recent arrivals on the scientific agenda is the origin and nature of consciousness. Clearly, a primary aspect of consciousness is feeling; our feelings, together with our thoughts, constitute the content of our awareness. Feelings can be about ourselves, such as when we experience pain, pleasure, well-being, or they can be about the outside world, as when we see a crying child, an injured animal, a dying tree. So within the question "Where does consciousness come from?" there is the question "Where do feelings come from?" The answer we are forced to give in science is that feelings arise from a particular dynamic organization of insentient matter, such as nervous systems at a particular level of complexity and order. Our feelings arise as emergent properties from something that has not the slightest trace of anything that could be called feeling or sentience. And here we face a problem.

The many examples we have of emergent properties in complex systems all have precursors of the emergent property in some form. For instance, the rhythmic behavior of ants tending the queen and brood in an ant colony can be described as an emergent property. This is because we cannot predict that this orderly behavior will arise from the activity of individual ants, which is actually chaotic, and their interactions, in which they excite one another. Nevertheless, rhythmic behavior is what is observed in real colonies, and it also occurs in computer models that simulate this behavior. This unexpected order consistently arises in systems organized dynamically in this way.

What is the dynamic precursor of the collective rhythm of ants in the brood chamber? It is the activity/inactivity pattern of individual ants. This pattern is chaotic in the technical sense of the term: There is no preferred periodicity. However, chaos is made up of a complex pattern of rhythmic components, so it is not hard to imagine that when ants interact by excitation, a preferred rhythm emerges. There is no miracle here of getting something from nothing. Nature is consistent, and once we see what happens, we can make sense of the phenomenon in terms of the behavior of the parts of a system and their pattern of interactions. This applies to the many examples of emergent behavior that occur in solid state physics as well as in biology.

However, if feelings emerge from matter that has not the slightest trace of what we call feeling, then we are indeed getting something from nothing. This sounds to me like a miracle. As a scientist, I prefer to put a tiny bit of feeling or sentience into matter in some form and allow it to get amplified in systems organized in particular ways—a view that has been extensively explored in the writings of such philosophers as Alfred North Whitehead (*Process and Reality*, 1929), Charles Hartshorne (*Whitehead's Philosophy*, 1972), and David Ray Griffin (*Unsnarling the World-Knot: Consciousness, Freedom, and the Mind-Body Problem*, 1998).

A Science of Quality

You can see where this leads. First, there is indeed feeling or sentience in matter, so animism is not so far off the mark.

But there is another aspect of science, just beginning to change, that carries this perspective on feelings and qualities much further. This change relates to the status of qualities. There is now evidence that when we look at an animal and conclude that it is nervous or boisterous or detached, we are observing an experience in the animal itself and not simply projecting our own feelings onto the animal. This evidence arises from studies carried out by the behavioral scientist Françoise Wemelsfelder and her colleagues showing that different people looking at the same animal have a high degree of consensus in their evaluations. Science is based on such consensus—consensus that leads to the conclusion that what is being observed has not simply subjective but real and objective status. What develops is a "science of qualities," a method of reaching consensus about such evaluations that the scientific community previously regarded as beyond the scientific pale.

As noted, our current science of quantities has given us the ability to produce enough goods to satisfy the needs of all of the planet's inhabitants, but it has left us with a rapidly declining quality of life worldwide. In the shadow of current science it is possible to see the components of a science of qualities which would restore qualitative evaluation to the place it occupies in our everyday lives, where judgments depend on quality as well as quantity. This restoration, together with the recognition that feelings belong not only to us but also to the rest of nature, in whatever form, presents us with a dramatically transformed set of possibilities for scientific knowledge, technology, and corporate and political action.

A shift in scientific perspective of this magnitude is not going to happen overnight, if it happens at all. It requires new forms of education at a basic level, in which the sciences and the arts are united to keep people whole and in which scientific and technological decision-making require participation by all members of civil society, with knowledge joined again to responsible action. Then the time we are living through now will be seen as a Dark Age indeed, but one in which the seeds of transformation were already present and lying within Earth's shadow, where Gaia was nurturing them, so to speak.

❑

BRIAN GOODWIN is a professor of biology at Schumacher College, Dartington, in Devon, U.K., where he coordinates a master's program in holistic science. He is also a member of the Santa Fe Institute. He is the author of *Temporal Organization in Cells; Analytical Physiology of Cells and Developing Organisms; How the Leopard Changed Its Spots: The Evolution of Complexity;* (with Gerry Webster) *Form and Transformation: Generative and Relational Principles in Biology;* and (with Ricard Solé) *Signs of Life: How Complexity Pervades Biology.*

MARC D. HAUSER

❑

Swappable Minds

CONSIDER THE FOLLOWING ODDITIES: A chicken with a piece of quail brain bows its head like a quail but crows like a chicken. A seventy-year-old man with Parkinson's disease, confined to his wheelchair, receives a piece of brain from a pig and in no time at all is out golfing, without a hint of his porcine accessory. This is not science fiction, à la Douglas Adams. This is scientific fact. Today we can swap brain tissue not just among individuals of the same species but between species. In the next fifty years such exquisite neurobiology will have revolutionized our understanding of the brain—of how it is wired up during development and how it has evolved over time. As we learn more and more about the brain, we will ultimately learn more about what it's like to be another species of animal. But the scientific and ethical consequences of this revolution are only just being contemplated.

What might it be like to be another organism? This question, formally articulated by the philosopher Thomas Nagel in his famous paper "What is it like to be a bat?" (*The Philosophical Review* LXXXIII, October 1974, 435–50), is generally about the mental lives of animals and specifically about their subjective experiences, their feelings. For some, it is simply not possible to recover such experiences, at least given current scientific tools. For others it is a difficult problem but within the reach of science. Since it is easier to talk about such issues if there is an example on the table, let me present one.

In the mid-1960s, two groups began experiments on rhesus monkeys designed to explore how they would respond to seeing another monkey receive an electric shock. At about the same time, the social psychologist Stanley Milgram began testing people to see how they responded to authority—in particular, whether they would obey an authority figure who instructed them to administer an electric shock to another human being. In one of the rhesus experiments, an individual was trained to pull levers to obtain his daily ration of food. When the monkey learned this task, a second monkey was introduced into an adjacent cage. Now when the first monkey pulled the levers, he delivered a severe shock to the other monkey. Surprisingly, not only did the first monkey stop pulling, but he did so for several days, even though by not operating the levers he forfeited his daily meal. He was starving, but the guy next door benefited by avoiding shock. Monkeys in control of the levers were more likely to abstain from pulling if the other monkey was a familiar cagemate than if he or she was an unfamiliar individual or a member of another species, such as a rabbit. Lastly, individuals who had been in the hot seat themselves and experienced a shock abstained longer than monkeys who had not had that experience.

The rhesus monkey experiments are particularly striking in light of Milgram's diametrically opposite results with humans, vividly described in his 1983 book *Obedience to Authority*. When an authority figure, such as an experimenter in a white lab coat, ordered a subject to pull a lever to shock another person, the subject repeatedly did so, even though the other person—an actor in no actual danger—reacted dramatically to the "shock." If a Martian descended to Earth to watch these two experiments, he would be forced to conclude that rhesus monkeys empathize while humans do not. Rhesus monkeys appear to know what it's like to be another in pain, while humans either don't or simply don't care. Of course we know that humans can care, can empathize with others, can consciously think about what it's like for someone else to have an emotional experience. Readers of George Eliot's poignant novel *Adam Bede* can readily feel what Adam feels when he sees Hetty, his unrequited love. The following passage should do the trick:

> That blush made [Adam's] heart beat with a new happiness. Hetty had never blushed at seeing him before. "I frightened you," he said, with a delicious sense that it didn't signify what he said, since Hetty seemed to feel as much as he did.

It is tempting to conclude that rhesus monkeys empathize and care about the well-being of others, and the experiments seem to support this point. But there are alternative interpretations. Perhaps the lever-pulling monkeys found the recipient's response to shock disagreeable; in aversive situations, individuals stop what they are doing. Perhaps the monkeys were concerned about retribution—that they might later be in the hot seat and at the mercy of a less than benevolent individual. If so, then abstaining from pulling the levers was not mediated by empathy but by self-interest. Regardless of which interpretation is correct, these experiments point to a particularly keen sensitivity to social context; they indicate that monkeys have emotions and goals and can act on them. We can use such information to make a connection between studies of animal minds and the promotion of animal welfare.

How might we find out about an animal's needs, desires, and goals? This is important information if we are to provide proper care for them. If only Hugh Lofting's Dr. Dolittle were nonfiction and, like him, we could talk to animals! There is, however, a fairly good substitute for such a conversation. It starts with careful observations of species-typical behavior and then borrows the tools of economics to ask what an individual will pay to obtain what it wants. Consider a recent study of farm-raised mink.

Mink farmers believe that their animals live in satisfactory conditions. What "satisfactory" means in this case is something like "has all the essential commodities for living a healthy life." Those who disagree with this view of mink life question the idea that "healthy" means sufficient food and water and no noticeable signs of ill-health. Armchair theorizing won't help this debate, but a crisp experiment will. The Cambridge University biologist Georgia Mason and her colleagues set minks up in individual cages that mimicked the conditions of mink farms: one nest box, drinking water, and food. Based on the assumption that all animals are pleasure-seekers, designed to obtain good things and avoid bad, each mink was also offered a choice between seven alternative compartments, each associated with some unique property: a water-filled pool, a raised platform, novel objects, a second nest site, a tunnel, toys, and extra space. To access each of these compartments, the mink was required to push open the corresponding door; on consecutive days, weights were attached to each of the seven doors, making it difficult to open. The experiment simulated a closed economy, wherein individuals are required to pay for what they want. The key intuition underlying these experiments is that animals may pay more not just for things they want but, crucially, things they need.

When the minks were released from their home cages, they consistently chose the compartment with the pool, spent the most time in this compartment, and paid the greatest costs to do so. Moreover, the levels of a stress hormone known as cortisol were measured and found to be as high when they were deprived of the pool as when they were deprived of food. What do minks want? Water pools. Why? Because in their natural habitat, they spend a considerable amount of time in the water, swimming and hunting for aquatic prey. Bottom line: In order to provide farm-bred minks with a "healthy life," mink farmers should spend the pittance it costs to buy small water pools for them. Minks without water pools are as stressed as minks deprived of food. And since no humane farmer would ever think of depriving them of food, why deprive them of a water pool? It makes no economic or ethical sense.

The rhesus monkey and mink experiments demonstrate how science can uncover what animals feel and want, and how such knowledge can be put to good practical use. But the techniques described are crude, given recent developments in genetics and the brain sciences. Now that we can alter an animal's genome by inserting or deleting genes as well as being able to remove or replace pieces of its brain, the range of possible questions we can ask and answer is vast. So is the range of potential ethical dilemmas. Consider the recent creation of smart, or "Doogie," mice (named after the precocious young TV character Doogie Howser). These mice were genetically engineered by the insertion of extra copies of a gene called NR2B, which plays an important role in memory formation. Mice with the extra genes were deemed smarter than controls, because they more rapidly learned to discriminate objects, respond to an aversive stimulus, and find a concealed ramp. Whether these increased abilities constitute the stuff of intelligence is certainly debatable; the results nonetheless showed a difference in performance that appeared to be mediated by gene manipulation. For those interested in the genetics of higher cognitive functioning, such results were quite stunning. They not only revealed the power of this technological advance but showcased the kinds of genetic engineering that might be used for applied purposes, especially the treatment of human medical disorders. By increasing the number of memory-related receptors in the brain, for example, one could theoretically reverse the devastating memory losses of Alzheimer's patients.

But the excitement about the findings on Doogie mice has been tempered by the results of another experiment— one that reveals the potential dangers of both gene and brain manipulation. Two years after the creation of Doogie mice, the scientific community was presented with an unanticipated by-product of, well, being smart. This by-product is best captured by the mantra of athletes: "No pain, no gain." Unlike their normal counterparts, Doogie mice have an increased awareness of acute pain for longer periods of time. This result has significant implications. As the geneticist Richard Lewontin has pointed out in *The Triple Helix: Gene, Organism, and Environment*, his critique of the Human Genome Project, we must avoid drawing naïve conclusions about the causal relationship between genes and behavior, lest we fail to appreciate the complex genomic and environmental contexts in which genes live.

It's a genetic jungle out there, and when a gene is removed, or replaced by another, or duplicated, we can only make educated—that is, statistical—guesses about the consequences. It's not that genetic or brain manipulations are worthless; on the contrary, such technologies are likely to open up a landscape of novel discoveries and insights. Along with such findings, however, we must be prepared to discover unpredicted complications and difficulties.

I frequently ask my students to conduct the following thought experiment: If you had the opportunity to undergo a reversible brain transplant (reversible because you can get the original parts back with no deficit), accepting some specific part of a willing animal donor, which part would you pick and from which species? Over the years, my students have placed the following three at the top of their lists: the olfactory bulb of a dog, the auditory cortex of a bat, and the visual circuits of an eagle. This thought experiment is equipped with a subtle trap. Although technology enables the insertion of these cortical regions, something else is needed to truly smell like a dog, hear like a bat, and see like an eagle. That something else is an interpretive system (along with such peripheral organs as the wonderful snout of a dog, the radar-dish ears of a bat, and the double fovea of an eagle's eyes). With a newly outfitted canine olfactory system, a human could detect millimoles of urine on a hydrant at a hundred yards but would interpret the odor as a human being does. The smell would probably be horrid, because of its intensity—a pungency that no human had ever experienced before.

I want to emphasize the importance of this interpretive aspect of our brain activity, because it is so often overlooked. A philosophical paradox and a horror film should help clarify the point. In logic, there is a theory of identity which states that for any two objects x and y with multiple parts, $x = y$ if every part of x is a part of y and every part of y is a part of x. The classic challenge to this notion of identity is the case of Theseus and his ship of Athenian sailors. When the ship sets sail, it is new. With time and wear and tear, the sailors replace the damaged planks with new ones. By the end of the voyage, all the original planks and fittings have been replaced. The paradox: Is the ship that ends the voyage the same as the one that started the voyage? Is it still the Ship of Theseus? Before answering, consider Roman Polanski's movie *The Tenant*, starring Polanski as a meek file clerk living in a Paris apartment. The previous tenant attempted suicide, and this launches the clerk into a state of delusional paranoia that leads to a rhetorical monolog involving the elements of the self: "If I cut off my arm, I say 'me and my arm,' but if I cut off my head do I say 'me and my head' or 'me and my body'?" These two cases make the difficulties associated with interpretation transparent. If we remove someone's olfactory system and replace it with a dog's or even another human's, we haven't changed that person's identity but only how the person senses odors (especially when the swap involves a dog); the person receiving this new circuit still places his or her own interpretive spin on the smell. However, when it comes to other brain parts, we must ask the identity question on a case-by-case level. As the neurologist Antonio Damasio has articulated in *The Feeling of What Happens*, his recent work on consciousness, different parts of the brain have different effects on the feeling of what happens to the self. Some swappable parts are likely to cause dramatic changes in identity, as the famous case of Phineas Gage makes clear. Gage, a hardworking and respected member of society, suffered damage to his frontal-lobe circuitry, an injury that transformed him into an unrecognizable individual, lacking in all moral judgment.

To take the "swappable minds" problem further, we can conduct another thought experiment—one that builds on some spectacular new results from the world of neuroscience. The neurobiologist Miguel Nicolelis and his colleagues have managed to record the electrical discharges of hundreds of neurons from an owl monkey's brain and use the signal to drive a robot's arm. This may sound like pure gadgetry, but it is not. It shows that at some level we can make sense of the neural code and understand how it mediates behavior. Now imagine that we could download the neuronal signals from any animal, creating a kind of hard-drive library of their thoughts while they were interacting with the world. We would be able to read the mind of an animal as it eats, sleeps, grooms, has sex, communicates. At some level we would have a deep sense of what it's like to be them. We would be a peeping *Homo sapiens*. We might even be able to match our own brain waves with theirs, thereby experiencing a kind of interspecies harmony never before achieved—clearly the ultimate in virtual reality games.

These are wonderful gedankenexperiments. Within the next fifty years the necessary technology will be available, though no one may choose to use it in such a fanciful way. The excitement lies in thinking about how much we will learn about the brain, both our own and those of other thinking creatures; the concern is that our technology is launching us into uncharted worlds with fuzzy moral consequences. If we swap brain parts, or turn genes on and off, who will be responsible for the consequences? The scientist? The doctor? The animal whose part was used to enable some human to have a better life? And if stem cell research

is approved, and different parts of the brain can be cultivated, should anyone be able to do a swap? For science to
profit from the creative energy of its contributors, the intellectual climate must support radical and even risky explorations, but scientists must also realize the potential ethical
consequences of their actions, and this includes studies of
nonhuman animals. As George Bernard Shaw muses in
Major Barbara, the secret of right and wrong "has puzzled
all the philosophers, baffled all the lawyers, muddled all
the men of business, and ruined most of the artists." He
could have added scientists, who must continue to struggle
with the distinction between the "is" of their results and
the "ought" of their conclusions.

❑

MARC D. HAUSER, a cognitive neuroscientist, is a professor
in the Department of Psychology and the Program in Neurosciences at Harvard, where he is also a fellow of the
Mind/Brain/Behavior Initiative. He is the author of *The
Evolution of Communication, The Design of Animal Communication* (with M. Konishi), and *Wild Minds: What Animals
Really Think.*

ALISON GOPNIK

❏

What Children Will Teach Scientists

IN 1997 SPACE SCIENTISTS working for NASA figured out how to tell whether there had once been water on Mars by analyzing the light reflected from Mars rocks. Water would leave traces of carbonate on the rocks and this would influence the spectrum of the light reflected from them. The scientists could work backward from data about the light, to the carbonates, to the presence of water. That year, a few miles away in a Berkeley preschool, a four-year-old boy named Kevin, with equal excitement, figured out how a new machine worked. When some combinations of blocks, but not others, were placed on the machine, the machine would play music. Kevin worked backward from this data to infer which blocks would make the machine go, and he used this knowledge to make the machine play music. In the next fifty years, we will come to understand how both Kevin and the rocket scientists at NASA could make these amazing discoveries. The answer to that question will change the way we think about science, childhood, brains, and maybe even genes.

Human beings know an enormous amount about the world around them. We know about rocks, waves, and toaster ovens; rabbits, palm trees, and potted petunias; parents, children, and orthodontists—an innumerable, inexhaustible array of objects, plants, animals, and people. By and large, our knowledge is strikingly accurate: We make remarkably good predictions about how toaster ovens and petunias and orthodontists work, and we use those predictions every day when we push the "Bake" button or add more Miracle-Gro or make an appointment. We didn't know all this when we were born; somehow or other we learned it.

We also learn about matters beyond our everyday experience—Mars rocks, viruses, neurons. And that knowledge, too, is strikingly accurate—accurate enough to let us conquer or at least alleviate such ancient scourges as smallpox and depression, not to mention baldness, impotence, and migraine headaches.

But how is it that we know so much? After all, the only information that reaches us directly from the world is a pattern of infinitesimal photons hitting our retinas and disturbances of air vibrating at our eardrums. How is it possible to get from that limited and apparently incoherent information to the truth? "Truth" may seem like a grand and metaphysical notion, but we all know a multitude of everyday truths: heat makes bread toast, water makes plants grow, broken appointments make orthodontists irritable. From a psychological point of view, our knowledge of these truths is as remarkable and puzzling as our knowledge of the truths of theoretical physics or astronomy. How is it that a series of interactions between one type of physical object, a bag of skin with a brain at the top, and other physical objects, like toaster ovens or petunias or orthodontists, can lead one object to learn about the other?

The last fifty years of developmental psychology have

made this question even more puzzling. New techniques have allowed us to understand more about children's minds than ever before. Babies and young children turn out to both know more and learn more than we would ever have thought possible. By the time they are three or four they have already learned fundamental facts about how the world works. A theory of learning has to explain how very small children, who can't yet read or write or even talk well, can learn so much so quickly. Our ability to learn can't just be due to education or training or elaborate social institutions—rather, it seems to be a fundamental part of our human nature.

In the past fifty years, cognitive science has told us a great deal about what our knowledge of the world is like, how we use that knowledge, and how that knowledge is encoded in our brains. Developmental cognitive science has also told us a lot about how our knowledge changes as we grow older. But we have not yet understood where that knowledge comes from or how it can give us a true picture of the world outside us. Learning has been relegated to the concluding "Unsolved Mysteries" chapter of cognitive science textbooks, right next to consciousness and romantic love. I'm not convinced that we'll understand consciousness much better in fifty years, and I'm even more skeptical about romantic love. But I do think that we will make real progress toward a scientific account of learning.

We can find a model for this account in another, apparently quite different area of cognitive science: human vision. Here is the problem of vision: Take the patterns of light that enter the eyes and turn that information into accurate representations of objects moving in space. How do we solve it? People seem to make some implicit and very general assumptions about how the light coming into their eyes is related to objects in space. For example, we seem to unconsciously assume that the light coming into the retina is a two-dimensional projection of a three-dimensional world, and we use this assumption to solve the problem of vision. We never think that we are living in Flatland, though logically we could be. In fact, babies seem to be born making this assumption; for instance, very young babies will pull back from an object that seems to be looming toward them.

But the really interesting thing is not so much that we know this fact about vision, but that assuming that this fact is true lets us discover an incredibly varied array of new facts. I make an unconscious general assumption that my retinal image is a 2-D projection of 3-D objects. This helps me to infer that the particular image on my retina at this moment must come from two disks connected by a thin rod, lying on the surface of the floor at a particularly weird angle. Knowing this fact in turn helps me solve the never-ending, ever-changing, practical challenge of finding my reading glasses.

Sometimes, of course, the assumptions may lead us astray, especially if a demonic psychologist has been at work inventing visual illusions. But much more often these assumptions are correct, and they let us draw the right conclusions about what the outside world is like.

But how can brains make assumptions? The assumptions I've been talking about translate into constraints on the output that a brain (or any other computer) produces when it receives certain inputs. When my retina fires in a particular way, only some neurons, and not others, will fire farther downstream. Neuroscientists can record the output of particular cells in the visual cortex while an animal looks at something and so construct a kind of circuit diagram. The neurological work shows how these constraints work in practice, and how these computations are actually carried out in the brain.

In vision science, there has been a remarkable convergence of different disciplines. Psychologists tell us what kinds of representations of objects we construct from what kinds of visual information; they tell us what we perceive when a particular pattern of light hits our eyes. That defines the problem. Mathematicians demonstrate how it is possible to solve that problem by making certain very general assumptions about how objects and light are related. Computer scientists show how those solutions can be implemented as constraints on the operation of actual physical machines. And neuroscientists show how those solutions are implemented in the particular machines inside our skulls.

A similar strategy may help us understand how we learn: that is, define the problems that human children and adults solve, mathematically work out possible solutions to those problems given certain assumptions, see how those solutions can be implemented in machines, and, eventually, see how they are implemented in our brains. Recently there has been a similar convergence of new ideas about learning from different disciplines—the philosophy of science, artificial intelligence, statistics, and developmental psychology. In the next fifty years, this convergence could lead to a full-fledged scientific theory of how we learn.

Start out with the problem, or at least one problem. How do we learn about the causal structure of the world— about how things work and how one event makes other events happen? This is obviously an important problem in the practice of any kind of science, but it is also an important problem for even very young children. Developmental psychologists have demonstrated that children understand a lot about causal relationships. By three or four, children understand some of the same basic causal facts about toaster ovens and petunias and people that adults understand. They also know more at five than they do at three and less than they do at seven; like scientists, children seem to be good at learning new causal facts.

But causal knowledge is also one of the most notorious examples of the gap between what we experience and what we learn. The philosopher David Hume originally articulated the problem. All we see are contingencies between events. One type of event may always follow another type of event, but how do we know that one event caused the other? And in real life, causal relations rarely involve just two events; dozens of different events may be causally interrelated in complicated ways. In real life, it's actually unusual for one event always to follow another; moreover, we may not always know which of two events came first. This uncertainty and complexity make even everyday causal problems complicated. Did the toaster-oven element smoke and burn the toast, or did the crumbs from the burning toast make the element smoke? Or did we set the temperature too high and independently burn the toast and cause the element to smoke? All we can see is the simultaneous mess.

Is there a way to sort out the mess? Intuitively, there are two things we can do. We can perform a series of experiments: For example, we could set the temperature knob up high without putting any bread in the oven, or we could scatter crumbs of burnt toast on the element while keeping the temperature low. If experiments are impossible, we can make careful observations to determine when the element smokes and when it doesn't. Does it smoke only when the temperature is high, whether or not there is burning toast in the oven? Or does it smoke only when there is burning toast, whether or not the temperature is high?

When we perform these experiments or make these observations, we are making assumptions about how the pattern of contingencies among a set of events is related to the causal relations among them, just as we make assumptions about how 2-D retinal images are related to 3-D objects. Just as we never think we live in Flatland, we never think we live in a world without causes. Of course, as with vision, a Humean demon could arrange the contingencies in a way that would fool us. But we progress by assuming that there are no such demons—that, to paraphrase Einstein, God is slick but not mean.

A group of philosophers of science at Carnegie-Mellon University led by Clark Glymour, and the computer scientist Judea Pearl and his colleagues at UCLA, have started to develop a mathematical formalism that allows us to go beyond intuition and state these assumptions in a rigorous way. We can think about causal relations in terms of a formalism called a directed acyclic graph, often also called a Bayes net. These graphs tell us about how the presence of one variable (like the state of the toast) will influence another variable (like the state of the heating element). The basic assumption behind the formalism is that if one event causes another, then when the value of one variable changes, the value of the other variable will also be likely to change. If the crumbs cause the element to smoke, then the presence of the crumbs should make the presence of the smoke more likely. We can represent these causal relations as arrows connecting the variables. The Bayes-net formalism makes some simple and general assumptions about how the pattern of causal relations—the pattern of arrows—

is related to the patterns of contingency among the variables. Here are three different graphical representations of the relation between the temperature knob, the burning toast, and the smoking element, corresponding to the three causal hypotheses we have described:

A. Temperature knob > burning toast > smoking element
   The temperature knob makes the toast burn, which makes the element smoke.
B. Temperature knob > smoking element > burning toast
   The temperature knob makes the element smoke, which makes the toast burn.
C. Smoking element < temperature knob > burning toast
   The temperature knob independently makes the element smoke and the toast burn.

Each of these causal structures has different implications for the patterns of contingency among the variables, given the basic assumptions about contingency and causality. This is what allows us to draw the right conclusions from our experiments and observations. For instance, if A is true and we remove the burning toast, we should no longer see any relation between the temperature knob and the element. If B or C is true, then we should still see such a relation. If we turn the temperature knob down and burn the toast independently, then the element will smoke if A is true but not if B or C is true. If we put the temperature on high but prevent the element from smoking, then the toast will burn if A or C is true but not if B is true. Similarly, different causal structures will lead us to observe different patterns of contingency among the variables, even if we don't do the experiments ourselves. The mathematical work lets us spell out all these connections between contingency and causation in detail, with structures much more complicated than those I've described here.

This mathematical work provides us with a kind of causal logic. Classical deductive logic took off from a few basic assumptions about reasoning and turned those assumptions mathematically into a method for deriving true conclusions from true premises. The new causal work makes a few basic assumptions about causality and then provides a systematic method for deriving true conclusions about causal relationships from observation and experiment.

Computer scientists have started to turn this abstract mathematics into computer programs that can actually learn about the world. One of the biggest differences between computers and people has been that computer programs could do only what you told them to do in the first place. A real Turing test—a test that would tell whether or not computers are like people—would require not only that a computer could do the same things as a human adult but also that it could learn how to do them based on the experiences of a human child.

Computer scientists translate the mathematical assumptions into constraints on the kinds of causal graphs a computer will produce when it is given certain patterns of contingency data. Using the new mathematical ideas, for example, computer scientists working for NASA designed programs that enable a robot to learn about the composition of rocks on Mars just by looking at data from a spectrometer, without having to consult experts back on Earth.

So far, all this may seem rather removed from the question we started out with: We wanted to know how ordinary people—and, in particular, ordinary children—actually learn, not just how high-powered scientists and statisticians and computers can learn. But there is beginning to be evidence that all learners may make the same mathematical assumptions about the relation between causality and contingency. Psychologists who explore the ways in which ordinary adults figure out causal problems have independently hit on some of the same mathematical models as the investigators in philosophy.

Psychologists are beginning to find that children as young as two use this sort of causal logic as well. We can present children with the equivalent of the toaster oven—a machine that makes things happen in a somewhat complicated and mysterious way. Sometimes we give the children particular patterns of evidence about the machine, and sometimes we let the children do experiments to find that evidence themselves. Then we see if they can figure out how the machine works. Children are surprisingly good at drawing just the right causal conclusions from the data in precisely the ways that the formalism would predict— young children really *are* rocket scientists. Of course, unlike scientists, the children seem to be completely unconscious of how they are reaching their conclusions.

In the next fifty years, once we know just what the computations are that children and adults perform, we will be able to look into their brains and see how they perform them. As our brain-imaging techniques become more and more precise and we learn more and more about the computations, we will begin to see how our brains are designed to implement those computations. The answer to that question is likely to be related to the greatest breakthroughs to come in neuroscience. The most important thing about the brain is its ability to change in response to input from the environment, yet this is one of the aspects of the brain we know least about. It is as if we knew everything about the anatomical structure of dead hearts but almost nothing about how living hearts pump blood. Brains are, above all else, organs that learn, and if we know how learning works we will know something important about how brains work.

What's more, as the fifty years proceed, the answer to the problem of how minds and brains learn may also turn out to be related to an even more general developmental problem, that of morphological development. Another great unsolved problem of the next millennium is the question of how DNA instructions turn something as simple as a fertilized egg into something as gloriously complex as a newborn baby. One of the lessons of the past few years of genetic research is that the genome can't be a detailed set of instructions for creating an organism; it isn't a blueprint. But then how does it work? Genes seem to operate not so much by directly determining what a cell will do as by initiating a kind of causal cascade in the environment of a cell that ends up influencing the cell in predictable ways. For example, the genes that determine sexual morphology do so by producing testosterone, which then acts on the organism in a complex way. Occasionally the demons are unleashed, the environment turns out to be different than the one the genes "expected," and the system goes awry. But usually the environment is predictable, and the genome exploits that predictability to produce a complex organism.

It may be helpful to think of DNA instructions as encoding implicit general assumptions about the interaction between cells and their environments (largely other cells), both before birth and after it. In the psychological case, implicit assumptions about our relations to our environment let us build complex structures that are remarkably well adapted to that environment. It is at least conceivable that this general approach may also apply to the biological case.

The greatest achievement of a unified theory of learning, though, may be to demonstrate that the most brilliant scientists and the most ordinary kids are engaged in the same enterprise. At the end of the last century, knowledge began to become the most valuable currency, like land in a feudal economy or capital in an industrial economy. The new science of learning should tell us that knowledge is not just a prize to be won in some desperate test-taking struggle for places in the contemporary mandarinate. Instead it is, literally and not just rhetorically, our universal human birthright.

❑

ALISON GOPNIK is a professor of psychology at the University of California at Berkeley. She is an international leader in the field of children's learning and was one of the first cognitive scientists to show how developmental psychology could help solve ancient philosophical problems. She is the coauthor (with Andrew Meltzoff) of *Words, Thoughts, and Theories*, and (with Patricia Kuhl and Andrew Meltzoff) of *The Scientist in the Crib: Minds, Brains, and How Children Learn*.

PAUL BLOOM

❑

Toward a Theory
of Moral Development

UNDERGRADUATES WHO TAKE their first psychology class—
typically a general survey course—are often surprised at
how dull it is. They come into the classroom with hopes
of learning about how their minds work, along with some
vague idea of what psychology is about—dreams, consciousness, evil, madness, love. At the end of the semester, they
stagger away with dim memories of inhibitory synapses,
Pavlov's dogs and Skinner's rats, some amusing and disturbing social psychology experiments, and the latest taxonomy of mental illness. But they leave without answers to
the questions that excited them—and, worse, with the
feeling that nobody is even asking such questions.

This is less true now than it was a decade ago, and there
are many signs that by the middle of the twenty-first century psychology will be anything but dull. It will be broad in
scope and rich in theory; it will bring together discoveries,
methods, and ideas from diverse fields, including evolutionary biology, cultural anthropology, and the philosophy of
mind. In other words, the psychology of fifty years from now
will look very much like the psychology we had over a hundred years ago, toward the end of the nineteenth century.

Now, those were exciting times. At the end of *On the Origin of Species*, published in 1859, Charles Darwin wrote: "In the distant future I see open fields for far more important researches. Psychology will be based on a new foundation, that of the necessary acquirement of each mental power and capacity by gradation." (Twenty years earlier, in his private notebooks, he had written, "He who understands baboon would do more towards metaphysics than Locke.") Darwin tried to make good on this promise in two subsequent books: *The Descent of Man, and Selection in Relation to Sex*, which was mostly an effort to explain the psychological difference between humans and other animals, and, a year later, *The Expression of the Emotions in Man and Animals*, which is still a profoundly important work for those interested in the psychology and physiology of emotional expression.

In 1890, William James published *The Principles of Psychology*, an ambitious and idiosyncratic view of the mind, summarizing the best science of its time. At the same time, from a very different direction, there is the work of Sigmund Freud. Despite Freud's continued influence in the humanities, he has an uneasy status within contemporary psychology, both clinical and experimental, and he might show up in the contemporary introduction course as a topic only of historical interest, if not as the target of ridicule. But his intellectual scope, enthusiasm, and ambition were extraordinary. As just one example, here are the first words of *Interpretation of Dreams*, published in 1899: "In the following pages, I shall provide proof that there is a psychological technique that allows us to interpret dreams, and that when this procedure is applied, every dream turns out to be a meaningful psychical formation which can be given an identifiable place in what goes on within us in our waking life." The spirit of Freud's broader research program—his goal of a unified science of the mind, based on the interaction of unconscious processes—was very much consistent with the work of Darwin and James.

In the hundred years that followed, psychology became insular, divorced from other fields, especially from philosophy and evolutionary biology. Psychology sought to become, intentionally and self-consciously, a science. One result of this attempt was the field of behaviorism, the dominant movement within American psychology in the twentieth century. Behaviorism, rejecting James's notion that psychology is the "science of mental life," assumed that only observable behavior can be objectively studied, that virtually all behavior is the result of learning, and that there is no principled difference in how different species (humans, say, as opposed to rats) do this learning.

Behaviorism is now as dead as any theory can be, largely because all of these premises have been proved wrong. And though there have been substantive methodological accomplishments that emerged from its research program—useful methods for testing the capacities of nonarticulate beasts such as rats and babies—few lasting discoveries relevant to the study of humans have emerged. The dominant movement in psychology today is cognitive psychology, which provides a computational analysis of mental life—most recently in terms of the dynamics of parallel distributed processing, or neural networks. This research program has been far more successful, but its success has been largely limited to just those areas that can be readily modeled on a computer. And so we have a huge amount of research on chess playing, deductive reasoning, object recognition, language comprehension, and different forms of memory. But the emotions, sexual behavior, motivation, personality, and the like have been largely relegated to more applied areas, such as clinical psychology.

This is all changing, mainly because of increased interaction with other fields. It is no accident that some of the most influential ideas in psychology have come from outside the field—from philosophers such as Daniel Dennett and Jerry Fodor, from evolutionary theorists such as William Hamilton and Robert Trivers, from economists, anthropologists, and linguists. Easily one of the most influential scholars in the field of psychology has been the linguist Noam Chomsky, whose 1959 attack on B. F. Skinner's book *Verbal Behavior* was a decisive blow against the behaviorist movement.

One particularly interesting cross-disciplinary connection has been with evolutionary biology. In the last several years, there has been a growing acceptance of the Darwinian notion that the brain, like any other biological organ, has evolved through the process of natural selection, and so the capacities of the brain can be profitably understood as adaptations and by-products of adaptations. This might sound obvious to some, and it is certainly uncontroversial within some fields of cognitive psychology. People who work on visual perception have never doubted that eyes have evolved for seeing, for instance. But in other fields of psychology, talk about evolution has been seen as unseemly, naïve, or politically suspect. When E. O. Wilson's *Sociobiology* came out in 1975, as a modern attempt to apply evolutionary thought to areas such as aggression, sexuality, and altruism, the reaction was decidedly hostile. But over the last decade the discipline of evolutionary psychology—a field that combines contemporary cognitive science with

evolutionary biology—has emerged, championed by scholars such as Leda Cosmides and John Tooby of the University of California at Santa Barbara.

Many of the specific proposals developed from this
framework have proved controversial, but the research
program itself is becoming increasingly accepted—so much
so that as a separate field of psychology it is likely to cease
to exist. In the next fifty years, "evolutionary psychology"
will be an anachronism, as the label suggests that there is
some other field of psychology that does not attend to considerations of selective advantage, adaptive design, and so
on. (Creationist psychology?) There will always be those
psychologists who are less interested in evolution, just as
there are those who are uninterested in how the brain works
or how development happens, but the fact that evolutionary considerations exist—along with those of neuroscience
and development—as a source of evidence in the study of
psychology will no longer be a matter of debate.

As a result of this reintegration with other domains, the
psychology of the next fifty years will be richer in many
senses. Traditionally, psychology has been the study of two
populations: university freshmen and white rats. But increasingly, researchers are becoming at least casually conversant
with research across species, not because of a naïve belief
that the mind is the same everywhere but as a way to look
at the evolution of mental systems. Similarly, one would
look at different points in the life span, at forms of mental
illness, cross-cultural differences, and so on, as a natural way
to explore the underlying structure and processes of mental life. This diversity of method is now standard for fields
dealing with memory and perception, but it will become
more common for the "softer" fields, such as social and personality psychology.

Take, as an example, the study of moral thought and action. This used to be a prime area of psychological inquiry. In fact, it was one point of difference between Darwin and James. In *The Descent of Man*, Darwin sought to explain human morality in terms of a general increase in human intelligence—one that enabled us to transcend the emotional reactions of our primate ancestors to appreciate the very notion of ethical behavior, of a code of morality that can be applied in a fair and objective manner. James had a different view, which he defended in *Principles:* that the unique aspects of human nature are merely the result of the addition of social instincts, such as shyness and secretiveness, that other animals lack. Darwin's contemporary, Alfred Russel Wallace, had a third view: He felt that human altruism was such a mystery that its very existence refuted the theory of natural selection as applying to the human mind. In general, the existence of altruism and morality, once investigated only by philosophers and theologians, has become the central problem of evolutionary psychology.

But the students in introductory psychology courses might hear nothing about this, instead devoting several hours to maze learning in the rat and the gross anatomy of the primate nervous system. My own field is developmental psychology, and it is striking that good textbooks in this area allocate more space to language learning—a field that has several journals and conferences specifically devoted to it—than to moral development: that is, how children come to have certain mature notions of right and wrong and how these notions come to affect their behavior.

There is no conspiracy here; the scanting of moral development is just because we know so little about it, not because of a lack of interest in studying it. In fact, the study of moral development is of profound practical importance.

Parents want to know how to raise their children to be good people; responsible members of society want to arrange the environment of the young—their schooling and so on—to raise a generation of good people. These desires appear to be universal, though of course people differ about precisely what counts as good and moral. And so we want to know the answer to specifics: Is spanking bad for children? What are the effects of violent video games or violent movies? Are children better off being raised by two parents than by one? How does day care affect children's temperament and empathy?

Despite what you might read in newspaper reports and television newsmagazines, we do not know the answers to any of these questions. There are some educated guesses, but basically all we know is that children who are raised by good parents are more likely to be good than those raised by bad parents. It doesn't matter much how one measures this: Whether "bad" as applied to parents means abusive behavior, alcoholism, schizophrenia, or poor attendance at PTA meetings, the sins of the parents often show up in the children. But we don't know why. Perhaps this phenomenon is an effect of parenting; being raised by aggressive adults might make a child more aggressive, for instance. Or a trait could be genetically transmitted, so that the relationship between the parents' aggression and the child's aggression would exist even if they had never met. It could be an effect of the child on its parents: A child who is aggressive is likely to provoke anger and violence in adult caretakers. There are other possibilities as well, and of course there is likely to be a complex interaction at work.

As I write this, a monograph has been published that reports a large-scale study of the effects of television viewing on 570 adolescents. Their television-watching habits were recorded at the age of five; then, ten years later, they were tested on grades, aggression, use of cigarettes, and similar indicators. The children who watched a lot of educational programs as preschoolers tended, as adolescents, to get higher grades, smoke less, and act less aggressively than those children who watched a lot of violent programs as preschoolers. The report brims with policy implications that accord nicely with common sense: educational programs good, violent programs bad, and so we should have more of the former and fewer of the latter.

But buried in the discussion section of the report is an admission by the authors that there is another explanation for their findings. We already know, after all, that some five-year-olds are more prone to aggression than others, some are timid, some like animals, others sports, and so on. The children in the study chose what television programs to watch, and presumably the bookish and intellectually engaged were more likely to watch *Sesame Street* and *Mister Rogers' Neighborhood*, while the more aggressive children tended to watch more violent shows. And so what the study may show is simply that aggressive preschoolers tend to become aggressive adolescents and bookish preschoolers tend to become bookish adolescents—something that, again, we have known for a very long time. Television might have nothing at all to do with it.

Or maybe television does have an effect. The point is that we just do not know. We do not simply need more and more studies; what we need is a theory of moral development, one informed by work across disciplines, including cognitive psychology and evolutionary theory. We need a theory of moral development that is at the level of intellectual richness of our theories of language development and perceptual development. Only then can we sensibly address these issues of causality and prevention.

Will we get such a theory in the next fifty years? I have been cheerful so far: Psychology will become more interesting, arbitrary borders between disciplines will dissolve, the scope of study will expand, and so on. This is all to the good. But it is an open question as to how much progress we will make on deeper problems, like moral thought or consciousness. There are those, such as Noam Chomsky and the philosopher Colin McGinn, who are skeptical. We are humans, not angels, after all, and just as there are some things that we can understand, there have to be other things that we cannot. It may be that the nature of moral thought or consciousness is simply beyond our understanding, not because they have a special, mystical status but because we aren't smart enough to understand such things. We might be like dogs trying to understand calculus.

There is no way to know whether this pessimistic view is right, and no alternative to moving along in the hope that it is wrong. But the notable lack of progress so far should lead to some humility on the part of psychologists, particularly with regard to the formation of social policy. This lack of progress is why I have not spoken about the practical benefits that psychology will yield over the next fifty years—an omission that may seem curious. Shouldn't a psychology of the future be expected to cure the mentally ill, fix unhappiness, get rid of prejudice and ignorance, show us how to raise moral, happy, independent children, and all sorts of other good stuff? That is the impression one receives from much of the popular press, and many psychologists, cheerfully confident in their abilities and eager for grant money and political clout, do much to encourage it.

In fact, the practical benefits of psychology have always been modest. If we put aside certain genuine clinical innovations—most of which have emerged primarily from biochemistry and neurology—psychological claims about how to manage society, treat criminals, and educate and raise our children have been, at their best, common sense. At their worst, they are faddish and dangerous—for example, the widely promoted claim within my own discipline that if you don't deluge a child with social and cognitive stimulation in the first three years of its life, you have lost the child forever. Similar examples include the benefits of playing Mozart to babies, the dangers of day care, and the crucial importance of mother-child bonding in the first hours of life. The only positive thing to say about these public pronouncements is that they change quickly. If you don't like what psychologists are telling you about how to raise your child—the amount and type of discipline, sleeping arrangements, and the like—just wait a year or two and they will tell you something different.

An optimistic view of psychology in the next fifty years is that it will be a more mature science, applying the methods and theoretical perspectives that have worked so well in domains such as perception to softer and less understood areas such as moral thought. In the course of this inquiry, we might develop enough of an appreciation of how the mind works, and enough confidence in our scientific approach, that we can acknowledge and appreciate how difficult these problems are and how much is left to learn.

❑

PAUL BLOOM is a professor of psychology at Yale University. He is an internationally recognized expert on language and development, and with Steven Pinker coauthored one of the seminal papers in the field. He is one of the youngest full professors at Yale and has published more than fifty chapters and journal articles in psychology, linguistics, cognitive science, and neuroscience. Bloom is the author of *How Children Learn the Meanings of Words* and the forthcoming *Bodies and Souls*.

GEOFFREY MILLER

❑

The Science of Subtlety

Too many of us have become stingy materialists when it comes to contemplating ourselves as emotional beings. A stingy materialist takes the view that subjective experiences may not be real if they have not yet been associated with particular brain areas or neurotransmitters or genes. They suppose that if we have found the brain area associated with a particular kind of pain, then that pain is a scientifically validated reality—but if we haven't yet found the brain area for sexual jealousy or existential dread, those emotions are probably not "real" and ought to be regarded with skepticism. Likewise, if we have found neurotransmitter deficits in schizophrenia, then it is a genuine medical disorder, but if we have not found such deficits in irritability, then perhaps that is not a real disorder but a mere character flaw or a bad habit.

Stingy materialists lack confidence not just in their own consciousness but in the future progress of their own materialist doctrine, with the result that they fetishize neuroscience, seeking its approval for all things subjective. Since neuroscience is still in its infancy, this overdependence on its current, limited efficacy results in an infantile view of human nature, in which people appear cartoonishly simple, portrayed in crude outline and primary colors.

Generous materialists like Charles Darwin and William James took quite a different view, confident that all subtleties of thought and feeling are based in brain activity. Believing equally in scientific materialism and human consciousness, they were generous in attributing rich subjective experience to the human species, with its complex brain. In the next fifty years, I predict that we will all become much more generous in our materialism, in the manner of Darwin and James. As neuroscience maps more of the subtleties of human consciousness, these subtleties will more readily be accepted and appreciated, to the benefit of our relationships and our society. Stingy materialism tends to make people selfish and arrogant, because only the cruder abilities and emotions that we share with animals tend to be taken seriously in neuroscience. Our more recently evolved, distinctively human capacities—for creativity, kindness, humor, imagination—remain understudied in brain-imaging labs. Generous materialism will make us more empathic and humble; we will realize that other minds and brains share all the subjective richness of our own experience.

The key to this revised and more benevolent view of human nature will be the development of new technologies for mapping neural activity and gene-activation patterns in the brain, which will show us exactly what is going on in our heads—not only when we do simple cognitive tasks that we have no particular feelings about but when we participate in complex social interactions that provoke our most human emotions. As the intricacies of our thoughts, feelings, and social interactions are objectively validated by our new technologies, science will develop a more nuanced view not only of the psychological complexities within each person but of the differences between people.

A century ago, we had to rely on the novels of Henry James to portray human consciousness in high-resolution detail and rich-spectrum color. In the future, we won't be able to rely on mass culture to do that—Viacom and Disney don't see the profit in it. But we may be able to turn to science to fill the void.

A little history can put stingy materialism in perspective. In the nineteenth century, psychology was practiced by and for the benefit of the Western intellectual bourgeois male. In this patriarchal psychology, little attention was paid to women, children, nonintellectuals, members of other cultures, or nonhuman animals as conscious beings. But this severely limited approach had an underappreciated advantage: The psychological similarities shared by Western intellectual bourgeois men allowed them to develop sophisticated ways to describe and communicate the subtleties of their rich inner lives. There was a positive feedback loop between bourgeois culture and bourgeois psychology, exemplified by the correspondence between William James and his brother Henry. They traded introspections fervently and got locked into a fraternal one-upmanship in trying to describe consciousness. More generally, the refinement of late-nineteenth-century European culture was reflected in the scope of psychological theorizing by such men as Charles Darwin, his cousin Francis Galton, Sigmund Freud, the philosopher Franz Brentano, the American psychologists James Mark Baldwin and William McDougall, and the German experimental psychologist Wilhelm Wundt. They weren't afraid to speculate on the nature of the emotions, aesthetics, love, family life, and even altered states.

This style of psychology all but disappeared in the twen-

tieth century, with the democratization of mass culture and
the rise of reductionism and positivism in science. Psychology expanded its subject pool—admitting to its precincts
women, children, the working class, non-Western peoples,
and primates—but narrowed its subject matter. At the same
time, the culture industry expanded its clientele, with mass
media such as film, radio, and television replacing the elite
intimacy of the novel and the theater. Cultural depictions
of human nature gradually became more stereotyped and
less nuanced for this mass audience. In parallel, there was a
gross simplification of psychology's content. The behaviorism of John Watson in the 1920s and B. F. Skinner in the
1950s viewed subjective experience as illusory, concen-
trated on learning as the foundation of behavior, and
excluded from psychology almost everything else—thinking, feeling, and social interaction.

With the cognitive revolution of the 1960s, computation
replaced learning as the dominant metaphor in psychology.
This liberated some psychologists to develop computer
models of perceiving, reasoning, and talking. Yet some modeling did not really broaden the discipline's subject matter
very much. It was still taboo for serious psychologists to
write about real social, sexual, or family relationships, or
such diverse states of consciousness as romantic love,
parental pride, or professional jealousy. Positivism, empiricism, and reductionism shifted the burden of proof onto
those who wanted to take human consciousness into
account; any state of mind that could not be validated in
the laboratory was regarded as unreal. Since almost nothing about a human being's conscious inner life could be
laid bare in this way, given the experimental methods then
available, most of subjective human life was excluded from
the science of psychology.

In short, both Western culture and Western psychology became Americanized in the twentieth century. They became more inclusive but less sophisticated, more objective in method but less accurate in result, more politically progressive but less humane. They also focused on individuals as atomized strangers, stripped of social, sexual, and familial relationships. Finally, they became more efficient at describing and exploiting simple knee-jerk reactions to advertising and propaganda but less accepting of any conscious state involving ambiguity, imagination, empathy, moral judgment, or aesthetic discrimination. Psychology assumed that it had to choose between analyzing behavior and understanding consciousness, between empirical respectability and subjective accuracy.

I think this assumption will be proved false in the next fifty years. New technologies have the potential to validate a much greater range of human subjective experience. The result may be a richer and more accurate model of human nature—but only if scientists work to study real thoughts and feelings in meaningful social situations.

For example, brain imaging methods can show which areas of the brain are active when we are doing particular mental tasks in a laboratory. Until now, most such tasks have been taken from standard perceptual or cognitive psychology and have no intrinsic meaning to the participants. I was a subject in one such brain imaging experiment as an undergraduate, and spent six hours at the Columbia Medical School comparing simple geometric shapes to each other while strapped to a bed, with twenty Geiger counters pointed at my brain, breathing a mix of oxygen and radioactive xenon gas. Certainly it is useful to know which parts of the brain are involved in shape discrimination. But while I was pondering the shapes I also had many thoughts and feelings unrelated to shape discrimination: worries about my last argument with my girlfriend, attempts to reconstruct the plot of a Pasolini film, speculations about President Reagan's apparent senility. From the viewpoint of the researcher, these passing thoughts were "noise," and given a sufficient diversity of such experimentally irrelevant states of mind, the researcher could be confident that they would all average out in the end.

However, there is something peculiar about experiments that pretend that the thoughts and feelings of the participants are much simpler than they really are. We should admit that current brain imaging technology is just not good enough for us to study fluid, complex thoughts and feelings about socially meaningful situations—and we should strive to develop systems that can. To some extent, better spatial and temporal resolution in our instrumentation will help; if we can measure precisely which cubic millimeters of the brain are active on a millisecond-by-millisecond basis, we will be able to study much more subtle psychological processes. A change in research style may also be necessary. We could revive the nineteenth-century tradition of introspection, as practiced by Franz Brentano and William James, in which psychologists were their own best subjects. We could put ourselves in the brain imagers and systematically explore our states of mind to see what lights up. But this would have the same behavioral limitation as armchair introspection: We cannot pay attention to empirical data about our brain states while simultaneously carrying on the sorts of real social interactions that are subjectively all-consuming, such as conversing, flirting, bargaining, arguing, breast-feeding, and so on. To do that, we will need brain imaging systems that are light, robust, mobile, and inconspicuous; only then will we be able to map the brain's true repertoire of capacities.

The other key technology in validating the complexities of human consciousness will be the mapping of gene-expression patterns. Each brain cell has a full set of genes, but only some of them are expressed at any given time— that is, only some are transcribed into RNA and thence into proteins. Moreover, different brain areas have different gene-expression patterns, and these gene-expression patterns change over time, not only across development from embryo to adult but also across situations from day to day and month to month. There are feedback loops between social environment, neurophysiology, gene expression, and behavior. When we fall in love, when a friend dies, when we get a major job promotion, no doubt the gene-expression map of our brain changes. Almost every state of mind lasting more than a few hours may involve changes in gene expression, and scientists have barely begun to track these changes.

Once the technology improves and we can track gene expression in real time, a new world of psychological complexity will open up. We will be able to see the correspondence between modern social situations and the genetically evolved behavioral capacities they elicit. We will also get beyond the muddied nonsense that nature and nurture are "inextricable," and see more clearly how specific situations, thoughts, and feelings activate specific genes—and vice versa. The charge that evolutionary psychology is a set of "just-so stories" will vanish, as we see the genetic footprints of evolution all over our brains.

If we have the courage to use them sympathetically, new developments in brain imaging and gene-expression mapping will illuminate a much broader range of human experience. If we can find objective neural and genetic markers in conscious states that now seem ephemeral or idiosyncratic, those states will be taken more seriously as a universal part of human nature. We shouldn't really need this objective validation, but we do: There seems to be an innate tendency to act as if one's own mental life were much more complex, meaningful, and valid than that of anyone else. The nineteenth-century introspectionists focused on themselves, forgot about others, and described a rich inner world. The twentieth-century behaviorists forgot about themselves, focused on others, and described a crude human nature based on learning and computation. Twenty-first-century psychologists will break down this distinction between self and others, between subjective and objective, by showing the neural and genetic signatures of even the most fleeting, whimsical, and ambivalent manifestations of human consciousness. The result should be a much more humane, inclusive science of people. My hope is that when freshmen take Psychology 101 in 2050 their response will be "Aha! So that's why we feel x when y happens!" rather than the all too common reaction now: "What does this have to do with real life, anyway?"

❑

GEOFFREY MILLER is an evolutionary psychologist at the University of New Mexico and author of *The Mating Mind: How Sexual Choice Shaped the Evolution of Human Nature.*

❏

The Future of Happiness

ONE ISSUE THAT WILL BECOME central in the next fifty years is how we shall use the ability to control the genetic makeup of the human species. In the past, our ancestors used crude methods of genetic selection to determine which kinds of children survived to reproductive age. Now we are being handed the dubious gift of reaching the same goal through the auspices of science.

Long before anyone suspected the existence of genes, farmers recognized that the traits of parents were passed down to the offspring, and thus they could improve the yield of pumpkins or the size of pigs by selectively breeding the best specimens with each other. It was then easy to apply this principle to human beings. Plato devotes a large part of the fifth book of his *Republic* to the question of how to apply the practices used to breed hunting dogs to producing rulers for the perfect State he envisions. In chapter 459, for instance, he writes:

> [T]he best of either sex should be united with the best as often, and the inferior with the inferior as seldom as possible; and . . . they should rear the offspring of the

one sort of union, but not of the other, if the flock is to
be maintained in first-rate condition. Now these goings
on must be a secret which the rulers only know, or there
will be a further danger of . . . rebellion.

Earlier, in chapter 415 of Book III, he writes, "And God
proclaims as a first principle to the rulers . . . that there is
nothing which they should so anxiously guard, or of which
they should be such good guardians, as of the purity of the
race." In fact, all known societies have practiced what in
retrospect we could label "eugenics" or "genetic engineering." These practices were often justified in terms that have
nothing to do with biology—such as religion or custom—
but presumably they were carried out because they were
seen as contributing to the survival of the group. It is useful
to remember that the idea of all persons having the right to
reproduce is a recent one; previous societies survived by
granting that privilege primarily to individuals who were
likely to produce above-average children.

Positive practices encouraged the mating of individuals
with desirable phenotypic traits—including health, strength,
and beauty—and material success, such as wealth or power.
Differential reproduction was achieved by various means:
The almost universal practice of obtaining a dowry or
brideswealth before marriage ensured that the future parents would have enough resources and kin support to bring
up children who would not become a burden to the community.

Negative practices discouraged reproduction among individuals with traits that a given society deemed undesirable.
Some of these were little more than natural tendencies:
For instance, poor, unhealthy individuals were less likely to
marry and have children. But other means were much more

active, ranging from castration to infanticide. Often a cultural practice that seemed to have an entirely different purpose might nevertheless have a substantial eugenic impact. For instance, the Russian Orthodox Church adopted the ritual of immersing naked newborn infants in cold water in order to infuse them with the grace of the Holy Ghost and protect their souls from eternal damnation. An incidental consequence of this practice was that less than healthy infants would not survive baptism, thus removing their genes from the gene pool. One can only speculate whether such rituals survived primarily because of the peace of mind they conferred on the devout or the genetic advantages they provided. Presumably they were overdetermined, in that both set of advantages supported their existence relative to alternatives open to the culture at the time.

Most of these practices were hit-or-miss, without any foundation in an understanding of how different traits are transmitted from one generation to the next. But this situation is about to change drastically in the coming decades. Currently two of the liveliest branches of the human sciences are behavioral genetics, which tries to ascertain the degree of inheritability of such behavioral traits as schizophrenia, propensity to divorce, political beliefs, and even happiness, and evolutionary psychology, which searches out the mechanisms by which these traits are selected and transmitted from one generation to the next. Both approaches assume that nature and nurture are implicated in shaping our behavior, thoughts, and emotions—although, contrary to the learning bias of the last century, they favor nature more.

This trend is bound to be magnified tremendously in the next half century as a result of advances in genetics. Although few important traits are likely to depend on the action of a single or even a few genes, some genetic engineers are confident that the era of "designer babies" is at hand. Even if their optimism is misplaced, it would be foolish to ignore the impending decisions we may soon confront. It is interesting that leading human geneticists, of whom my colleagues and I interviewed close to one hundred in a recent study, have rarely taken the more controversial aspects of their work seriously. Most argue that it has no relation to anything resembling eugenics. They scoff at the possibility of human cloning and see little likelihood that their discoveries can be misused. Almost unanimously, they claim no special knowledge about or responsibility for potential applications of genetic engineering; they insist that this is a political decision, to be taken by society at large—even though "society" lacks the specialized understanding to make informed decisions. The situation is not unlike what was happening in atomic physics a little over a half century ago, when even such a universal thinker as Niels Bohr maintained into the 1940s that experiments with nuclear fission could have no possible practical applications.

But ready or not, the choices will soon have to be made, and they will determine our future. For instance, let us suppose that it will be possible soon to substantially increase g, the general-intelligence factor that underlies the linguistic and mathematical skills prized by the educational system and useful in other spheres of life as well. Is this a good idea? Several commentators have pointed out that society is already stratified by intelligence to a troublesome extent. Whereas in the recent past people could be considered successful if they were hardworking, honest, friendly, or virtuous, without necessarily being "book smart," nowadays abstract reasoning skills are becoming a prerequisite for any kind of material or social success. If we find ways to enhance this trait genetically, the trend may become exponential. As the division between "supersmart" and average individuals increases, so will the gap between their economic and political power. Endogamy based on intelligence—already in effect—will become more pronounced, as no one with an IQ over 200 would dream of marrying someone with an IQ of less than 150. If the engineering affects the germline, these divisions will be transmitted automatically to the next generation.

But what if, in an unlikely burst of egalitarianism, we found ways to enhance everyone's intelligence—to raise the baseline for the entire human race? Would that be a good idea? The answer is that we don't know. Most biological and psychological functions that are useful in small doses are dangerous when they become excessive. As Aristotle noted, virtues become vices when taken to the extreme: Courage turns into foolhardiness, prudence into indecisiveness. The ambiguous relation of genius to insanity suggests that too much intelligence may have its own handicaps— excessive sensitivity, for example, leading to proneness to anxiety and depression. Or, to the extent that rational intelligence is linked to self-centered attitudes à la Ayn Rand, it may result in a species even more unfeeling and cruel than we are now.

A more basic issue is whether, having the means, we should aim for uniformity or diversity in fiddling with the human genome. The pressure for uniformity is going to be great: Everybody will want to have children who are intelligent, good-looking (by standard conceptions of beauty), ambitious, and successful. Diversity is risky. Who would want to wager on the unknown, the untested? Yet the biologist E. O. Wilson's arguments in favor of biodiversity also apply to psychological traits; the prospect of an increasingly homogeneous race is not only frightening to our humanistic sensibilities but potentially dangerous from the strict perspective of survival. Because the future is largely unforeseeable, the best strategy is to have a diverse pool of potentialities from which adaptive responses to new situations may emerge, instead of locking ourselves into a pattern that is best in terms of present conditions.

If human genetic engineering will be market-driven (instead of being dictated by a central computer that will determine how many warriors, workers, and drones society will need in the next generation), it is likely that the most intense selective pressure will be for producing happy children. When parents are asked what they hope for their children, the typical answer is that they hope the kids will be well educated and have good jobs, but above all else that they will be happy in whatever path they choose for themselves. Contemporary parents seem to agree with Aristotle, in that they understand that while every other good is a means to an end, happiness is *the* good in itself: It is what we hope to achieve through education, money, beauty, and intelligence. If it becomes possible to produce happiness through genetic manipulation, that may well become parents' first priority.

According to behavioral geneticists studying identical and fraternal twins reared together and apart, at least 50 percent of happiness is genetically inherited. One might have some justifiable reservations about how "happiness" is measured in such studies, but the fact that there is a set point of happiness different from one individual to the next and relatively impervious to external ups and downs seems well established. Of course, the general level of happiness in a population is also affected by economic conditions (having more money is related to happiness, up to a point, but past the threshold of income that would be average in Portugal or South Korea additional income does not correspond to more happiness), the political situation, and many other external variables. Nevertheless, one's genetic inheritance plays an important role.

So let us suppose that in the decades ahead it will be possible to enhance the likelihood of our children's happiness through genetic engineering. Are we going to do them a favor by availing ourselves of this opportunity? Will society, and the species as a whole, benefit from such a choice? In speculating on what the answers to these questions may be, we might start by reviewing the little we know about happiness at this point.

In the first place, it seems clear that people's self-report of how happy they are is a fairly valid measure of their happiness. It correlates highly with the perception of family and friends, with the incidence of pathologies and relevant behaviors—in short, people who think they are happy also look and act like happy people are supposed to. They tend to be extroverted, they have stable relationships, they live healthy and productive lives. So far, so good.

But there might be some interesting downsides as well. For instance, one of the most widely accepted definitions of happiness is that it is a state in which one does not desire anything else. Happy people tend not to value material possessions highly, are less affected by advertising and propaganda, are not as driven by desire for power and achievement. Why would they? They are happy already, right? The prospect of a society of happy people should be enough to send shivers down the spine of our productive system, built on ever-escalating consumption, on never-satisfied desire.

Will academic psychology be of any help in providing answers to these impending choices? Until about two decades ago, the discipline had very little to say about happiness. It was considered too "soft" an issue for serious scientific study. To make a difference in this quandary, psychology will have to focus once again on its original object, the psyche—not as an ephemeral, mystical, soul-like substance but as a set of the very concrete phenomena that transpire in our consciousness as our attention is turned to apprehending, integrating, and responding both to external stimuli and to internal states (that is, thoughts and emotions). The stream of consciousness is considered by most scientists, including psychologists, to be too subjective for rigorous study, while in fact it is the most objective datum we have access to. Scientific facts and the knowledge based on them is hearsay that I am glad to accept on faith, but the events in consciousness, such as fear, joy, anger, hope—to them I have immediate access and their reality is beyond question.

For my part, I determined to develop a systematic phenomenology that would find answers to the following kinds of questions: How do people's thoughts, feelings, goals, and actions fluctuate during an average day? During a lifetime? How are these components of the stream of consciousness related to each other? When do people feel happy in everyday life? Any one of these questions could in turn generate dozens of further ones, including investigations of how age, gender, ethnicity, and other such differences affect consciousness and how patterns measured at one time relate to patterns measured years later. Among the things we learned is that people who are engaged in challenging activities with clear goals tend to be happier than those who lead relaxing, pleasurable lives. The less one works just for oneself, the larger the scope of one's relationships and commitments, the happier a person is likely to be.

It is also important to realize that consciousness has its own specific reality, which is immediately destroyed when one begins to analyze it in terms appropriate to less complex systems. For one thing, it is an open system, whose states constantly change through time. What's on my mind now, for instance, cannot have been accurately predicted by what was on my mind a minute ago, even if you had all the information about my brain chemistry, genetic background, past learning, and so forth, sixty seconds earlier. What happens between time 1 and time 2 is that any sound, sight, feeling, or idea that enters consciousness during that minute may set my thoughts and feelings on an entirely new and unpredictable course.

This indeterminacy can be seen most clearly in creative activity. It is generally thought that the elements of a poem (or a sonata, a painting, a scientific theory) could be retrieved from the poet's mind if we had enough information about the contents of that mind. That is, in a distant analogy to the homuncular theory of embryonic development, we believe that the creative work is contained—even if only in some microscopic or codified form—within the creator. But that is not the case. A poet may start with a single word or phrase—a word or phrase that is meaningless or ordinary but which at that particular moment seems compelling to him. Why the word or phrase is suddenly meaningful might be explained if you knew what the poet was thinking or feeling just before. But what happens next is not: The word may suggest ideas and associations that were not predictable, and these in turn open up new directions of thought and feeling, which lead to more words, and so on in an expanding circle of meaning that is the result of an emerging, autonomous, self-organizing system—still based on the poet's past consciousness but no longer reducible to it.

One need not turn to creativity to illustrate this process. Let's take a more universal event, the reaction of parents to their newborn child. Genetic and evolutionary psychology can tell us a great deal about how and why parents bond with their offspring. Parenting is one of the oldest human experiences; it has been the experience of every generation since the beginnings of our species. Nevertheless, even if one knows everything about babies and birth, seeing one's own child for the first time is an event so *sui generis* that nothing can adequately prepare one for it. Its nuances depend on how one feels about one's spouse, one's financial situation, one's life in general—to say nothing of the baby's physical appearance and behavior—and all of these elements are striving to achieve meaningful combination with the main event, the birth of the baby. You can guess what that combination will look like by knowing as much as possible about the parent, but the prediction will be imprecise, because too many of the variable factors that affect the parent's consciousness are external.

If psychology were to take the stream of consciousness for its territory, it might begin to provide the kind of knowledge that we will need to make enlightened choices about the sort of future we want. With every increase in knowledge, our responsibilities increase. In the past, we were like passengers on the slow coach of evolution. Now evolution is more like a rocket hurtling through space, and we are no longer passengers but its pilots. What kind of human beings are we going to create? Flesh-and-blood copies of our machines and computers? Or beings with a consciousness open to the cosmos, organisms that are joyfully evolving in unprecedented directions?

Psychology is beginning to show signs of moving in the latter direction. At various centers in the United States and abroad, topics like wisdom, life goals, intrinsic motivation, spirituality—all of which would have been outside the pale a few decades ago—are being investigated by serious scholars. During his recent presidency of the American Psychological Association, Martin E. P. Seligman established within the profession a "Positive Psychology" movement, which reaches beyond the traditional goals of healing mental afflictions. Among its accomplishments so far has been the development of a list of "strengths" that are ubiquitous across times and cultures—such as wisdom, valor, perseverance, and integrity. As a next step, the knowledge of how such strengths are cultivated is being assembled. Eventually this knowledge should permeate the profession, giving it equal weight with the practice of therapy and prevention. We will need such a science to confront successfully the challenges of the next fifty years.

❏

MIHALY CSIKSZENTMIHALYI is a Hungarian-born polymath, formerly chairman of the Psychology Department at the University of Chicago and currently Davidson Professor of Management at the Claremont Graduate University in Claremont, California. His research and theories in the psychology of optimal experience have been put into practice by such national leaders as Bill Clinton and Tony Blair, as well as the chief executive officers of many of the world's major corporations. His books include the best-selling *Flow: The Psychology of Optimal Experience; The Evolving Self: A Psychology for the Third Millennium; Creativity;* and *Finding Flow.*

ROBERT M. SAPOLSKY

❑

Will We Still Be Sad
Fifty Years from Now?

W HILE WE'RE ON THE first lap of this new century, there are two irresistible temptations: One is summing up—awarding the palm to the most significant event or achievement of the century just past; the other is looking forward. For someone in my business, the task is to choose the disease of the twentieth century and try to discern its trajectory in the next.

In contemplating candidates, one might focus on diseases that have been defeated since 1900. The logical choice in that area is smallpox—its elimination is one of medicine's great triumphs—but the sentimental favorite would have to be polio, despite the inconvenient fact that it is still rampant in many parts of the developing world. Here in the West, many still remember the iron-lung terror of the earlier half of the century, when polio was still a scourge. Many survivors in their fifties or older are laid low by post-polio syndrome, the last echo of that disease, in which neuro-muscular systems weakened but not destroyed by polio have aged for decades, producing muscle weakness and atrophy. Moreover, there's the human-interest angle of the Sabin/Salk contest. Or should one's choice rest on a disease that flourishes at roughly the same rate as it has throughout history—newsworthy for its very imperviousness to modern science? In that case, malaria is a contender. Then there are diseases that have risen to a notorious prominence in the past century—AIDS, obviously, along with cancer, heart disease, adult-onset diabetes, and Alzheimer's.

But if the prize is to go to a disease that devastates, remains astonishingly resistant to modern medicine, and has become epidemic, my nominee would be major depression.

By "major depression" I do not mean a setback that leaves us feeling miserable for days, before we realize that whatever it was is not the end of the world. Major depression incapacitates its sufferers for months, for years; they are sunk in a well of despair, hardly able to work, to love, to socialize, to sleep, to eat. They may even be unable to go on; approximately half of the people with major depression attempt suicide at some point in their lives. Depression is the textbook modern psychiatric disease: It is a biological disorder with genetic, neurochemical, and hormonal facets giving rise to mental "illness," and it is a disorder profoundly sensitive to an environment that produces feelings of helplessness.

Major depression is heartbreakingly common, afflicting about 15 percent of the people in the developed world at one point or another during their lifetimes. And it is becoming more common: The rates of depression in Western countries have steadily climbed during the last fifty years. While some might question this finding as potentially spurious—since depressed people today are more likely to seek medical help than in past times, and health care professionals are more likely to diagnose depression than were doctors in the 1950s—these studies are among the most rigorous epidemiological studies ever done in psychiatry and carefully controlled to account for such confounds. The rate of depression is indeed ever increasing.

And how do I think major depression will fare in the next fifty years? Unfortunately, I suspect that this medical disaster is not about to disappear and could well become more prevalent.

Why this conclusion? To begin with, it is critical to understand the connections between stress and depression, and also the particular ways in which our lives have become more stressful. We (and our bodies) are more likely to consider some external challenge unbearably stressful if we lack a sense of control over it. Or if we lack predictive information as to when it is coming and how bad it's going to be. Or if we lack social support, including outlets for the frustrations that arise as a result of it. An extraordinarily useful model of depression was developed in the 1970s by the psychologist Martin Seligman and his colleagues at the University of Pennsylvania. It is called "learned helplessness," and builds upon these variables. Confronted by a psychological stressor, most of us can put it in context, put a perimeter around it, realize that this stressor is not the whole world: We cope. Depression is a failure of that perimeter, a globalizing of the experience, producing the distorted conclusion that "not only is this development something awful that I have no control over: *Everything* is awful, and there is nothing in my life that I have any control over." The sufferer has learned to be helpless. While stress of a sufficient severity pushes virtually anyone to that conclusion, individuals who are biologically at risk for depression have a lower threshold. On some biological level, major depression is a failure to re-equilibrate after a stressor, instead succumbing to a permeating sense of helplessness which then takes on a life of its own.

Why have we become depressed at an increasing rate and why do I think that rate will continue to rise? With any luck, there is much that should buffer this epidemic. A young girl, say, fifty years from now will likely have more control over who she becomes as an adult—a neurosurgeon, a CEO, a soccer star—than have young women in the past. Institutionalized segregation, Jewish quotas, "No Irish Need Apply" signs will be obscure history. Many of the traditionally peripheralized will be getting a somewhat less severe training in helplessness. And in some ways, our planet might actually become less brutal: One can optimistically predict an increasing percentage of people living with at least some semblance of self-government. There might well be decreases in the worldwide rates of slavery, widow-burning, and rape—although perhaps that is expecting too much of our species.

Moreover, science has found increasingly ingenious ways to combat this disease. We've learned about some of the neurotransmitters—chemical messengers in the brain—that are probably out of whack in depression. The most notable example is serotonin, which has a vast number of functions, including roles in emotional regulation that may well have something to do with depression. At present, the best guess is that depression involves either too little serotonin being used as a messenger or too little sensitivity of target neurons to that serotonin message. The strongest evidence for this idea is the activity of the renowned antidepressant Prozac, which fairly selectively increases the amount of serotonin available for signaling between neurons. The next-generation Prozac is nearly here; it will be faster and stronger, with fewer of the side effects of its predecessor, such as occasional sexual dysfunction in male patients and problems with memory or concentration. We've also learned something of how stress, a sense of helplessness, and some of the hormones that are secreted in such situations can give rise to some of the neurochemical changes of depression. As a result, novel therapies are now being pioneered to treat depression at the hormonal level.

Other bits and pieces of our knowledge of depression are falling into place, too. In the brains of many long-term depressive patients there are areas that seem to be abnormally small—in particular, the hippocampus, which plays a key role in the formation of certain types of memory. People with histories of long-term and severe depression have the sort of memory deficits that would be associated with such hippocampal atrophy. And we are learning something about the genetic contributions to vulnerability to depression—genetic differences in the working, say, of serotonin trafficking in the brain or of stress-hormone synthesis.

Such findings are bound to open up new therapeutic vistas, and hints of these are already in the works. Perhaps the most exciting progress will occur in the increasing understanding of the biology of what makes us individuals. Out of that knowledge will come a greater understanding of the individual triggers of depression.

Given our scientific advances and the factors that are reducing our sense of helplessness, why do I assume that we will be getting sadder? Mainly because it strikes me that there is still so much in our present civilization that is depressogenic. I can already imagine a certain exasperated critique of where I'm heading. "You're about to whine about how tough things are now? How about Barbara Tuchman's Europe? Or the Great Depression? Ever hear of World War II?"

From what I know of fourteenth-century life in Tuchman's distant mirror, it seems perfectly plausible to me that everyone was depressed back then, that the psyche of medieval Westerners was fundamentally different from ours. But in contemporary times, the challenging features of the last fifty years are particularly prone to generating depression—largely because the depressing features of life occur, with increasing frequency, outside the context of social support. Our traditional sources of solace will doubtless progressively atrophy in the years to come. "Family" will have to accommodate a divorce rate unlikely to decrease; the connectiveness of stable communities will have to accommodate our prized freedoms of mobility and anonymity. Nowadays it is the rare person in our culture who spends a lifetime living in a small town surrounded by relatives and friends.

Moreover, our technology isn't likely to help reduce our stress, despite (or maybe even because of) our expectation that it will. We will continue to come up with inventions that save us time, and then, as usual, we will readjust upward our expectation of how much there is to get done. We will fashion more material luxuries, but then we will recalibrate our baseline sense of entitlement. We have our zillions of gadgets and our leisure-filled lifestyles, but often these options are empty ones, as we struggle to decide which breakfast cereal, plastic surgeon, new-model car, or new-model spouse will be the one that finally makes us happy.

Many of our stressors are tailor-made for a globalized sense of helplessness. While there may be more restrictions on brutality in our perhaps more civilized future, those who violate the restrictions will have a greater technological reach, the town bully with a cudgel wielding instead an arsenal, or a militia. And whatever our global media village metastasizes into from its current mere five hundred cable channels, it will allow us to marinate in the visceral immediacy of every horror in its purview: the drive-by shooting in the next town, the genocide on the next continent, the tragedy of a dying child or a dying ecosystem.

But there is a single critical and statistical reason why depression rates seem likely to grow. As we try to rebound from our fin-de-siècle glaze—a time when we learned that ethnicities can be cleansed, high schools can become slaughterhouses, and First Families can be sordidly dysfunctional—our children are sitting beside us. Two vital facts: The rising incidence of the rate of depression is most pronounced among adolescents and young adults; and major stressors early in life reliably increase the risk of depression in adulthood. Childhood is a time when one learns the range and efficacy of one's control over one's external circumstances and what sources of comfort can be relied upon. And we have allowed our children to learn at ever younger ages a toxic secret—that the world is full of pain and sadness and there's not a damn thing that can be done about a lot of it. No child can be as practiced as an adult at constructing perimeters of containment in the face of this discovery. One response might be to develop a detached sense of entitlement, whereby living well is the best revenge; this tendency has become an epidemic in its own right. The other response, more prevalent among those who combine empathy with reflection on their own human condition, is a lowered threshold for learning to despair. The seeds of this despair have already been sown in the next generation.

But what about better living through chemistry—the next pill that will make us better than well and safe from pain? Currently available antidepressants aren't very effective, overall. Many depressive patients have to stop taking their drugs because of the intolerable side effects. Other sufferers must experiment with different drugs for years before they get some relief; many never do. But why not assume that there will eventually be a knockout punch, the pharmaceutical breakthrough of Depression Begone that will consign this disorder to the archives? Evidently it's not going to happen and scientists know it. In the course of writing this piece, I tried to find a marketing vice president from a drug company who would supply me with a quote along the lines of "Depression? Piece of cake—this one's beat in twenty years." But no one would. Even the people who are hired to be optimistic about the conquering of disease were making no promises about this one.

And that's no surprise. Medical progress can come in a variety of ways—draining swamps to get rid of mosquitoes, tracking down the world's last case of smallpox in some African village, understanding how a cell becomes cancerous or how some Machiavellian virus destroys the immune system that was meant to destroy it. But depression is a medical problem that humbles those approaches; there will never be a vaccine against the vicissitudes of life. Ultimately the question that one must ask of the scientists researching this disease, of the clinicians, even of the evolutionary psychologists, is not why so many of us will succumb to depression but how most of us will manage to avoid it.

❑

ROBERT M. SAPOLSKY is a professor of biological sciences at Stanford University and of neurology at Stanford's School of Medicine. He is also a research associate at the National Museums of Kenya. While his primary research, on stress and neurological disease, is in the laboratory, for twenty-three years he has made annual trips to the Serengeti of East Africa to study a population of wild baboons and the relationship between personality and patterns of stress-related disease in these animals. His latest book, *A Primate's Memoir,* grew out of the years spent in Africa. He is also the author of *Stress, the Aging Brain, and the Mechanisms of Neuron Death,* and two books for nonscientists, *The Trouble with Testosterone and Other Essays on the Biology of the Human Predicament* and *Why Zebras Don't Get Ulcers: A Guide to Stress, Stress-Related Diseases, and Coping.*

❏

Fermi's "Little Discovery" and the Future of Chaos and Complexity Theory

ON DECEMBER 2, 1942, in secret experiments conducted in a squash court at the University of Chicago, Enrico Fermi created the first self-sustaining chain reaction, a crucial step in the development of the atomic bomb. That is how he will always be remembered, at least by the general public. But among scientists, he is revered for his remarkable breadth. Fermi was perhaps the last person at the top of his field as both a theorist and an experimentalist. "He struck me as the cleverest man I had ever set eyes on—well, perhaps the cleverest man with one exception," was Jacob Bronowski's tantalizing assessment. "He was compact, small, powerful, penetrating, very sporty, and always with the direction in which he was going as clear in his mind as if he could see to the very bottom of things."

Just before he died in 1954, Fermi had some fun playing with what physicists would call a toy problem. It was a simple, beautiful question—not meant to be anything too realistic, just a way to explore a fundamental issue that he'd always wondered about. Now was his chance. Fermi was visiting Los Alamos, and he was looking for any excuse to

test-drive the new MANIAC, the world's first supercomputer. It was like a sports car he couldn't resist.

Working with John Pasta and Stanislaw Ulam, Fermi
asked the machine to simulate thousands of vibrations of
an elastic chain of thirty-two particles. The whole system
was supposed to represent an idealized, one-dimensional
lattice of atoms held together by chemical bonds. For small
vibrations, chemical bonds behave linearly: Stretch them
twice as far and they pull back twice as hard. All of traditional solid-state physics is built on that approximation.
But Fermi knew that real bonds would act nonlinearly if
the vibrations became large, and no one had a clue what
would happen in that case. The mathematics available at
the time could not provide the answer; no one could solve
the equations for a nonlinear system of so many particles.

Of course, that was exactly the point. Fermi concocted
this problem because it was impenetrable by conventional
methods. Now, with the help of the MANIAC, he and his
collaborators were about to shine a spotlight on nonlinearity, the darkest corner of classical physics. What they discovered was shocking. They expected that when they
disturbed the chain from its rest state, the nonlinearities
would eventually cause the system to thermalize—that is,
to degenerate into a state of randomness, with all possible
modes of vibration jiggling equally strongly. That's what
thermodynamics said would happen. But the computer
said no. After a very long time, the particles returned almost
exactly to their starting positions. This bizarre echo was
humanity's first indication that nonlinearity could be a
source of astonishing order. Nonlinearity giveth chaos, and
nonlinearity taketh it away.

Fermi was delighted by the echo phenomenon—accord-

ing to Ulam, he referred to it affectionately as "a little discovery"—but unfortunately he died before he could see the
results published. Pasta and Ulam never felt comfortable
writing the joint paper without him; instead, they quietly
circulated it as an internal Los Alamos report and waited
ten years before finally publishing it as part of Fermi's col-
lected works.

The Fermi-Pasta-Ulam problem must have looked particularly odd back in the early 1950s. In those days, physics
was all about quantum electrodynamics. Nobody was thinking about something as hoary and desiccated as classical
mechanics. Hadn't that been thoroughly picked over for
three hundred years? Fermi recognized that, on the contrary, the subject had barely been touched; all the nonlinear problems were as baffling as ever. In retrospect, we can
see how farsighted Fermi was, both in his choice of problem and in the revolutionary computer experiment he
invented to tackle it. His vision—properly generalized—is
actually more appropriate to our own era and to the next
fifty years of nonlinear dynamics.

In 1953, nonlinear dynamics had barely mastered questions
involving two coupled oscillators, let alone much larger
numbers of them. Oscillators were the province of engineers. Throughout the first half of the twentieth century,
the engineers built wonderful devices that exploited nonlinearity: vacuum tubes that powered the earliest radios
and televisions; phase-locked loops for radar and commu-
nications; lasers for precision optics and eye surgery. All
these inventions relied on self-sustaining nonlinear oscillators—specifically, on their tendency to synchronize with
each other or with an incoming signal. But all these technologies used small numbers of oscillators, usually just one

or two. Enormous arrays were out of the question; the mathematics for predicting the collective behavior of a large number of these units did not yet exist.

The only discipline that could handle vast numbers of interacting particles was statistical mechanics, the branch of physics originally designed to account for the behavior of gases composed of trillions of molecules. Fermi was a master of statistical mechanics, and he knew full well that it worked beautifully for systems at thermodynamic equilibrium. Unfortunately, nonequilibrium phenomena were another matter altogether. And that was precisely the stunning outcome of the Fermi-Pasta-Ulam simulations: The system did not settle down to equilibrium in the expected way. Ordinary statistical mechanics was out of its depth here.

In the fifty years since Fermi's experiment, nonlinear dynamics and statistical mechanics have both matured—and to some extent overlapped. Several of the great theoretical triumphs of the past few decades harnessed techniques from both disciplines to make striking advances. The physicist Mitchell Feigenbaum used the renormalization group, a Nobel Prize–winning method from statistical physics, to show that there are certain universal laws governing the transition from regular to chaotic behavior. His predictions were quickly confirmed in systems as diverse as heart cells, chemical reactions, and semiconductors. The theoretical biologist Arthur Winfree showed that the synchronization of huge networks of biological oscillators was strikingly reminiscent of a phase transition, like the sudden freezing of water into ice below a critical temperature. Other seminal models of complex systems—Stuart Kauffman's model of gene networks, Per Bak's self-organized sandpiles, John Hopfield's artificial neural networks—have all been illuminated by the fusion of statistical mechanics with nonlinear dynamics.

The progress of nonlinear dynamics follows a simple logical structure, governed by a few organizing principles. The most important one is that smaller systems are easier than larger ones. The first kinds of nonlinear problems to be understood were those involving just two variables—for instance, a swinging pendulum, whose state is completely characterized by its current position and velocity; given those two numbers, we can predict its exact location at any moment in the future.

Problems with three variables can be much wilder: They can be chaotic. Chaos means that a system governed by deterministic rules can nevertheless behave in random and seemingly unpredictable ways. With the work of Edward Lorenz and other chaos theorists from about 1960 to 1985, the pervasiveness of this erratic form of behavior was recognized and its universal features explained. It cropped up everywhere, from population fluctuations in ecosystems to the irregular dripping of a faucet. Soon chaos was being tamed for practical purposes, like encryption and even the composition of musical variations.

The frontier of nonlinear dynamics has since moved on to much larger systems, typically networks composed of enormous numbers of interacting units. In that respect, the Fermi-Pasta-Ulam problem, with its large array of coupled oscillators, has a contemporary feel. Within the class of such coupled systems, there are further organizing principles that guide us to the most tractable problems. Some of these are statements about the individual components of the network: For instance, the collective behavior of oscillatory units is easier to predict than that of chaotic ones. Identical units are easier than diverse ones. Other principles deal with how the units are connected: Networks with regular or random architecture are easier than those with more elaborate connectivity. Taken together, these heuristics point us toward the study of large systems of identical oscillators coupled in discrete lattices or other simple arrangements, and that is how nonlinear dynamics has evolved in the past few years. Some of the hottest topics are synchronization in arrays of many coupled lasers, neurons, or the superconducting devices known as Josephson junctions. Instead of discrete lattices, some researchers are looking at pattern formation in continuous media, like fluids, chemical reactions, and nerve and heart tissue. Perhaps the most exciting work has to do with the dynamics of spiral waves, thought to be implicated in ventricular fibrillation, the most pernicious form of cardiac arrhythmia.

Close cousins of these large, nonlinear systems are what researchers at the Santa Fe Institute have called "complex adaptive systems"—imaginary worlds consisting of millions of competing organisms, chemicals, companies, or stock traders, each adapting to their environment and thereby altering the environment for everyone else. The computer-simulated models of these situations are clearly speculative, but they offer intriguing clues about the impact of natural selection on a stunning variety of problems: the resilience of ecosystems, the chemical origins of life, the battle of firms in the marketplace, and the booms and crashes of the stock market.

In many ways, these Darwinian simulations are the intellectual grandchildren of the Fermi-Pasta-Ulam study. Notice the resemblance in how they treat the computer: not as a number-cruncher but as an exploratory tool, a flashlight to help us grope in the darkness. Notice the shared fascination with unexpected order in complex nonlinear systems.

And notice the lack of realism in the models. Just as Fermi's one-dimensional chain of particles was a deliberate simplification of a crystal lattice, today's computer-simulated complex adaptive systems are crude caricatures of real ecosystems and real markets. For now, that's good strategy, but in the next few decades the field has to grow up. The challenge is to find ways to incorporate more reality without sacrificing insight.

The first hurdle is to characterize the connectivity of complex networks. Instead of the idealized random or regular topologies assumed in our existing models, we need to learn how real networks are actually structured. That's essential if we are ever to understand how the brain computes or why cells become cancerous. In the past three years, we've started to explore the detailed architecture of networks ranging from food webs and nervous systems to power grids and the Internet. The surprise is that despite their bewildering diversity, they share some universal structural motifs.

For example, they all display the small-world phenomenon (popularly known as six degrees of separation): Nearly all pairs of nodes are connected by remarkably short chains of links. Furthermore, the number of connections per node tends to follow a power-law distribution, with a much heavier tail than the normal bell curve. That means that the overwhelming majority of nodes are poorly connected but there are also a significant number of gigantic hubs—like Yahoo on the Web, or ATP in biochemical-reaction networks. These topological features must influence the kinds of collective dynamics the networks can support—their resistance to random failure or deliberate attack, their ability to propagate contagion or support global computation, and so on. But at the moment we have no idea how to relate a network's topology to its overall dynamics.

In fact, we have few good models of dynamics at all, outside the physical sciences. Although networks pervade biology, sociology, and economics, we know very little about the rules governing the interactions between genes, people, or companies. That is the second hurdle that must be cleared in the next fifty years. Our models of complex systems will never advance beyond caricatures until we can find a way to infer local dynamics from data.

A classic example of how to do that is Alan Hodgkin and Andrew Huxley's reconstruction of neural dynamics from their electrophysiological measurements of the squid axon (1952). They had the benefit of being able to perform controlled experiments, in which the voltage across the nerve membrane could be adjusted to any value and then held there. By measuring the flow of sodium, potassium, and other currents through the membrane as a function of the voltage, they derived a precise portrait of the nonlinear dynamics of a single neuron.

The trouble, of course, is that this strategy is not general enough. In other settings it may be impossible to clamp any variable at a series of desired levels. Still, there may be more indirect ways to make the necessary inferences. For instance, consider the genetic networks that control the workings of a cell. With the DNA chips now available, we can measure the simultaneous activity of thousands of different genes as a function of time, but we still don't know which genes are talking to which and how they are influencing each other's activity in a quantitative way. All that information is somehow reflected in the data on the DNA chips, but we don't know how to crack the code. If we can develop systematic ways to infer dynamics from multiple time-series measurements, that advance will have tremendous implications—not just for biology but for sociology and economics as well.

The problem of characterizing connectivity is much easier than the problem of inferring dynamics. Yet even if we manage to clear both hurdles, we are going to run smack into the most fundamental obstacle of all. It is inescapable once we commit ourselves to exploring nonlinear systems with millions of interacting variables. It arose in embryonic form in the Fermi-Pasta-Ulam problem, and the truth is, no one knows what to do about it. We've been ignoring it or conveniently changing the questions we ask, but at some point it has to be confronted directly.

The difficulty, simply put, is that our brains can visualize only three dimensions. Evolution has hardwired us that way. With computer-assisted training, maybe we could dimly perceive a few more, but I doubt we'll ever be able to picture millions.

Why does this matter? Because for a hundred years, since Henri Poincaré discovered chaos in the gravitational three-body problem, geometry has been our best ally in nonlinear dynamics. Keep in mind that nonlinear equations generally cannot be solved in closed form, so algebraic formulas are out. But Poincaré showed that we didn't need formulas. By drawing the right kinds of pictures, one could understand many of the key qualitative features of a nonlinear system. Poincaré's method assigns one axis to each state variable, so when just two or three variables are involved, we can easily visualize the dynamics. But for today's problems with millions of variables, we're stuck.

The geometric approach is still valuable, and some progress has been made by using more abstract forms of reasoning, but without direct visualization we are dynamically blind.

That's why turbulence has defied theoretical understanding for so long. Even though we've known the governing equations for over a century, we can't see how their solutions behave in the "state space" that Poincaré defined. In particular, we can't picture the attractors—the essence of the long-term dynamics—because the state space is infinite-dimensional.

You could argue that we've slain this beast before. In the statistical mechanics of gases or magnets, the state space has Avogadro's number of dimensions (a 23-digit number), yet we still understand those systems very well. True, but only if they are at thermodynamic equilibrium. Then we know the statistical properties of the long-term behavior; that's what the equilibrium distribution tells us. And that's precisely why we are in trouble now. We don't understand the long-term statistical behavior of complex systems because they are far from equilibrium. And since we can't visualize the attractors either, we don't know what to do.

As Richard Feynman put it forty years ago, "The next great era of awakening of human intellect may well produce a method of understanding the qualitative content of equations. Today we cannot. Today we cannot see that the water-flow equations contain such things as the barber-pole structure of turbulence that one sees between rotating cylinders. Today we cannot see whether Schrödinger's equation contains frogs, musical composers, or morality—or whether it does not. We cannot say whether something beyond it like God is needed, or not. And so we can all hold strong opinions either way."

If we're ever going to reach that next great era of awakening, we'll need to be rescued from the devil of dimensionality. Look for computers to be our saviors. Once they become reasonably intelligent, they should be able to visualize any number of dimensions. They already do the grunt work of running our simulations; maybe the day will come when they will also extract the laws of self-organization in complex systems.

That speculation raises a broader issue: Will we still enjoy doing theoretical science when computers become better at it than we are? If they phrase their insights in qualitative terms that we can grasp, the computers will seem like prostheses—mere extensions of ourselves, no more threatening philosophically than electron microscopes. Otherwise, if they leave us in the dark, they'll seem like oracles, inscrutable and often disturbing. In parts of mathematics, that's already happening. Some theorems have been proved by computers, but because the proofs involve such numerous or intricate subcases, no human can check them. Some of the chess moves that Deep Blue made against Garry Kasparov had this same character.

I wonder if that's what the future holds for research on complex systems. We may end up as bystanders, unable to follow along with the machines we've built, flabbergasted by their startling conclusions.

Enrico Fermi may have been the first person to feel this eerie sensation. His invention of the computer experiment created a whole new way of doing science. It seems only fitting that it was made possible by the work of his contemporary John von Neumann, the architect of the earliest high-speed computers and a man that Jacob Bronowski once described as "the cleverest man I ever knew, without exception."

❏

STEVEN STROGATZ is a professor in the Center for Applied Mathematics at Cornell University. He is the author of the best-selling textbook *Nonlinear Dynamics and Chaos: With Applications to Physics, Biology, Chemistry, and Engineering* and the forthcoming trade book *Sync*. His seminal research on human sleep and circadian rhythms, scroll waves, coupled oscillators, synchronous fireflies, Josephson junctions, and small-world networks has been featured in many publications and broadcast outlets, including *Nature, Science, Scientific American*, the *New York Times, The New Yorker*, BBC Radio, and CBS News.

STUART KAUFFMAN

❑

What Is Life?

IN THIS TRIUMPHANT ERA of molecular biology and the first draft of the human genome, one might have supposed that we would know the answer to the question, What is life? Yet we do not. We know bits and pieces of molecular machinery, patches of metabolic circuitry, genetic network circuitry, means of membrane biosynthesis, but what makes a free-living cell alive escapes us. The core remains mysterious.

I suspect that in the next fifty years we will answer this question, and that the answer will require a marked change in physics and chemistry, to say nothing of biology. Indeed, for biology, an understanding of the fundamentals of what life is will set the stage for a general biology, freed of the constraints of terrestrial biology, the only biology we know. We will be able to ask whether there are laws that govern biospheres anywhere in the universe.

It is not that fine minds have failed to attempt an answer. Perhaps the most famous is a book with the same title as this essay, published in the war-heavy year of 1944 by no less a giant than the physicist Erwin Schrödinger. Yet Schrödinger's brilliant book took as its central question one whose answer was not likely to answer the question his title, and mine, poses. His central question concerned the source of the stunning order in biological systems. In answer, Schrödinger reasoned that that order could not be due to statistical averaging, where one would expect fluctuations on the scale of the square root of the number of particles. Using recent X-ray induction rates for mutations, Schrödinger correctly realized that a gene could be made up of at most a few hundred to a few thousand atoms. Square-root-n fluctuations would not fit with the heritability shown by organisms. So Schrödinger made some brilliant leaps: He argued that order required the stability of chemical bonds—in particular, covalent bonds—which depend upon quantum physics, not classical physics. Then he noted that a simple crystal cannot "say" very much, because once the structure of the unit crystal is known, the identical repeats of that unit add no ability to "say" more. So Schrödinger shrewdly put his bet on aperiodic crystals, whose detailed structure would contain a microcode that specified the development of the organism. He was right— only nine years would elapse until Watson and Crick discovered the structure of the aperiodic solid of DNA, and another decade or so until the microcode, in the sense of the genetic code, was understood.

But if Schrödinger brilliantly foresaw the source of order in organisms, did he answer his question, What is life? I think not; I cannot say all at once why I am so persuaded. This essay is part of my effort to explain an alternative attempt at the question.

I begin with a different image and question. Picture a bacterium swimming up a glucose gradient. We all unhesitatingly say of the bacterium, without attributing consciousness to it, that it is going to get food. That is, the bacterium

is acting on its own behalf in an environment. I will call a
system that can act on its own behalf in an environment an
autonomous agent. All free-living cells and organisms are
autonomous agents. But the bacterium is "just" a physical
system of molecules arranged in some way. So my question
becomes not, What is the source of order in biology? but,
What must a physical system be to be an autonomous
agent?

I will leap ahead to give you my tentative answer: I
believe an autonomous agent must be a physical system
capable of self-reproduction and also capable of performing at least one thermodynamic work cycle. Take the bacterium, its flagellar motor rotating and doing work against
the aqueous medium in which it is swimming. The bacterium is able to reproduce and is, in swimming, doing
work cycles.

A number of issues immediately crowd upward for
attention. In describing the bacterium as acting on its own
behalf, I am in fact applying language that we use in
describing the actions of human beings. With respect to
humans, the concepts of "doing," "acting," "purpose," and
"choice" are all familiar, deeply embedded parts of a
Wittgensteinian language game, or way of life, that we all
live. Even were we to doubt that it is appropriate to apply
the concept of "acting on its own behalf" to a bacterium,
the troubling question of whence comes the language of
action, doing, and purpose applied to humans in a world of
physics would arise. We, too, are just physical systems. So
let's jump over the philsophical problem of other minds,
and the problem of how far into the biological world we
want to extend our language game. My own sense is that
we do in fact extend that language to free-living organisms,
even single-celled bacteria, to say nothing of a pair of birds

building a nest or your dog chasing the stick he has cajoled you into throwing. So grant me my question about the bacterium. If I am right, autonomous agents are the necessary and sufficient condition for the language game of action and doing, hence acting on their own behalf.

It may be that in my tentative definition of an autonomous agent, I have inadvertently stumbled across an adequate definition of what life is. Certainly agency seems coextensive with life, even given the inevitable boundary cases—mules and the like—that are clearly alive but cannot reproduce. I will not insist upon the adequacy of my definition of autonomous agents, nor that that definition suffices for life—but I suspect that it does.

A brief word about definitional circles in science. Consider Newton's famous $f = ma$, and ask how you can define f independent of m. Force is that which accelerates mass; inertial mass is that which resists acceleration by force. Poincaré held the view (with which I am sympathetic, but not all physicists are) that $f = ma$ is a definitional circle, with force and mass defined circularly in terms of one another. This definitional circle does not prevent celestial mechanics from being the triumph that it is. Take Darwin's claim that natural selection picks out adaptations and that adaptations are features that lead organisms to be more likely to produce offspring. Again, here is a definitional circle that does not prevent evolutionary theory from using Darwin's insights with considerable success.

My definition of an autonomous agent is simultaneously a definitional circle of a kind and a jump to a new language game—that of doing and acting. It is not truly a definitional circle, in the sense that "self-reproduction" and "work cycles" are defined independently of autonomous agents, but the further identification of autonomous agents with the ability to act on their own behalf is a definitional circle. Again, that need not mean that the definition is without merit or scientific usefulness. As I have struggled to understand and unpack what is contained in my brief definition of an autonomous agent, I have been led into strange byways of conceptual analysis that suggest the definition is at least rich and provocative. Others must say whether ultimately it will prove useful.

Here is a sketch of a plausible chemical system that is both self-reproducing and carries out a work cycle. The system rests, first of all, on a portion that is able to reproduce itself. This could be either a single strand of DNA 6 nucleotides long which is able to catalyze the ligation of two 3-nucleotide-long sequences that, once joined, turn out to be identical to the initial DNA hexamer; or a 31-amino-acid sequence able to catalyze the ligation of two fragments—one 15 amino acids long, the other 17 amino acids long—into a copy of the initial 31-amino-acid sequence. By the way, the success of the Ghadiri group at Scripps in getting a protein to self-reproduce shows once and for all that molecular replication need not be based on the template replication of a DNA- or RNA-like molecule.

The work-cycle part of my hypothetical autonomous agent is achieved with some imaginary chemistry in which pyrophosphate, PP, breaks down to two monophosphates, P+P, with a loss of free energy. That free energy is coupled into the excess replication of the DNA hexamer or the 31-amino-acid sequence. Here, "excess" means that more of the reproduced molecule is produced than would be produced were that reproductive reaction not coupled to the source of free energy given by the pyrophosphate. Once the pyrophosphate is broken down, I imagine a further free energy source—an electron that absorbs a photon, exciting the electron. As the electron falls back to its unexcited state, this source of free energy is used to drive the resynthesis of pyrophosphate to levels beyond the equilibrium concentration it would achieve in the absence of the electron's energy kick. The existence of a work cycle is established by the fact that the phosphates in the *PP to P+P then back to PP* reaction cycle "rotate" around that reaction cycle rather than reaching equilibrium. It is a molecular motor performing a thermodynamic work cycle.

A number of features of my tentative autonomous agent stand out. First, the system works only if displaced from chemical equilibrium; agency is a nonequilibrium concept. Second, the system is a new class of perfectly legitimate nonequilibrium chemical-reaction networks, ones that couple self-reproduction and work cycles. We just have never linked the two before. But we can investigate them experimentally now. Third, the physicist Philip Anderson has pointed out to me that by the excess synthesis of the DNA hexamer or the 31-amino-acid sequence, the total system has stored energy that could later be used to correct mistakes, just as happens in DNA repair enzymes in your cells.

Up to this point, this essay is legitimate science. From here on, things will become successively stranger and perhaps more problematic.

The first issue is to examine the concept of "work" in more detail. For a physicist, work is just force acting through a distance. Work is done in accelerating a hockey puck with a hockey stick, and is just the sum of the little *ma* terms that constitute that accelerated motion.

But is work so simple after all? In any particular case of work, the force is applied in an "organized" way to achieve the work. Physicists bury the consideration of the organization of process in their initial and boundary conditions.

But where did the initial and boundary conditions come from? This is typically not answered by the physicists, and from the point of view of analyzing the motion of a cannon-ball shot from a cannon, it suffices to ignore previous events and take the initial conditions of the cannon and its elevation, powder, shot weight, wind condition, and so forth and calculate forward. But it took work to make the cannon, load in the powder and shot, and so on. And somehow the full becoming of the event of the cannon firing reaches backward to the Big Bang. A biosphere is all about the coming into existence of ever novel initial and boundary conditions over the past 3.5 billion years, and the physicists' isolation of a part of the system "now" will not suffice to answer the question of the evolutionary emergence of such persistently novel initial and boundary conditions.

A second way to begin to realize that there is something incomplete about the concept of work is to realize that an isolated thermodynamic system—say, an enclosed gas in a thermodynamically isolated box—can do no work. But if the box is divided into two parts by a membrane, then one part can do work on the other part; for example, if the gas pressure is higher in the first part, the membrane will bulge into the second part, doing mechanical work on it. Thus work cannot be achieved in the universe unless the universe is divided into at least two regions. Furthermore, just where did the membrane come from?

The definition of "work" I find most congenial comes from Peter Atkins's book on the second law of thermodynamics, wherein he points out that work is a "thing"—namely, the constrained release of energy. Consider a cylinder with a piston inside and a compressed working gas between the piston and the cylinder head. The gas can expand, doing work on the piston, pushing it down the cylinder. What are the constraints? Evidently the cylinder, the piston, and the location of the piston inside the cylinder, with the gas trapped between the two. But where did those constraints come from? Well, it took work to make the cylinder, work to make the piston, and work to put the gas into the cylinder and the piston in afterward. Thus we come to an interesting new cycle, not previously noted. It appears to take work to make constraints and constraints to make work!

Indeed, when one tries to unpack Schrödinger's concept of information contained in the microcode of his aperiodic crystal, the "saying" of something comes to have a physical consequence only by the arrangement of constraints on the release of energy, which, when released, constitutes work. But there is more, for the released energy that does work can be used to construct more constraints on the release of energy, which constitutes more work, which in turn constructs more constraints. Note that these notions are not in the physics or chemistry we have been taught. One begins to have the sneaking hunch that all this constraint construction on the release of energy—which, as work, can construct more constraints on the release of energy—has something profound to do with an adequate theory of the organization of processes. We have as yet not even the outlines of such a theory, and that outline is certainly not contained in information theory.

Nor is the point I am making merely rhetorical. A dividing cell does precisely what I just said. For example, the cell carries out thermodynamic work to construct lipid molecules from fatty acids and other building blocks. The lipids then can fall to a low-energy structure—a bilipid layer that forms a hollow "bubble" called a liposome. Indeed, cell membranes are just such a bilipid layer that forms a hollow ball. Now the membrane itself is used by the cell to alter constraints. Consider a hypothetical pair of small organic molecules, *A* and *B*, which can undergo three distinct chemical reactions:

1) *A* and *B* can react to form C and *D*;
2) *A* and *B* can react to form E;
3) *A* and *B* can react to form F and G.

Each of these reactions has reaction coordinates. Thus, as *A* and *B* react to form C and *D*, chemical-potential walls and barriers guide the de-forming *A* and *B* molecules into the C and *D* product molecules. Similar walls and barriers guide the reaction to form E, and F and G. Now the chemical-potential walls and barriers constitute constraints on the reactions of *A* and *B*. Furthermore, let *A* and *B* dissolve from the aqueous interior of the cell into the cell membrane. Once in the membrane, the translational, rotational, and vibrational modes of motion of the *A* and *B* molecules alter. This in turn changes the chemical-potential walls and the height of the barrier separating *A* and *B* from C and *D* (or from E, or from F and G). Thus, the formation of the cell membrane, which took thermodynamic work, has altered the constraints on the chemical reactions of which *A* and *B* are capable. Further, the cell does thermodynamic work to link together amino acids into a protein enzyme that happens to bind *A* and *B* and convert them to C and *D*, rather than to E or to F and G. Thus the cell does thermodynamic work to construct constraints on the release of (here chemical) energy down particular pathways. And once released, that energy can do work to construct further constraints.

What I have said is correct, but not normally talked about. Indeed, the cell carries out what I have come to call "propagating work"—work that links the organized release of energy at one point to the building of constraints and the organized release of energy at other points, summing up into a closure of process, by means of which the cell builds a rough copy of itself. This organization of process is carried out by any dividing cell, yet it is stunning that we have no language—at least, no mathematical language of which I am aware—able to describe the closure of process that propagates as a cell makes two, makes four, makes a colony and, ultimately, a biosphere. This self-propagating organization of process is contained in the concept of an autonomous agent. Note that while what I have said violates no law of physics, physics and chemistry appear to lack the language to discuss it. Note, too, how rich the tentative definition of "autonomous agent" begins to seem. Somehow, in a way that I can only intuit, the cell exhibits a form of organization that is not captured by our concept of information—a concept that leaves out any mention of constructing constraints on the actual occurrence of anything in the real physical world. Indeed, the proliferating coevolution of the biosphere is a story of the persistent arising of ever more subtle and refined ways in which free energy is captured and released in constrained ways in which new constraints and structures are built and the work propagates to extract work and build yet more subtle structures. Just think of a rain forest. It arose with no central authority in charge and has created a burgeoning, interwoven diversity of linked processes. Yes, the biosphere creates propagating linked patterns of organized work. But I don't even begin to know how to mathematize the concept.

The reason I don't know how to mathematize the concept may point to the most odd, frustrating, and potentially profound aspect of this investigation: The biosphere may actually be doing something that cannot be stated at all beforehand. If so, the way Newton, Einstein, Bohr, and Boltzmann taught us to do science is limited. The biosphere may persistently alter its "phase space." I know of no mathematical framework that can describe this process. I would also note that if my claim is true, it wreaks havoc on the frequency interpretation of probability, where one can in fact state all the possibilities beforehand.

The issue ties in with what Darwin termed *preadaptations*. But first, plain adaptations. What is the function of your heart? Well, the function of your heart is to pump blood. But your heart also makes heart sounds, which are not the function of your heart. So the function of your heart is a subset of its causal consequences. Darwin would say that the causal consequence of your heart for which it was selected was to pump blood, not to make heart sounds. So the simple part of this is that to know the function of *A* and *B* above, or any part of an organism, you must know the whole organism in its entire environment. Autonomous agents have an inalienable holism about them.

Now to preadaptations. Darwin's idea was that a part of an organism might have causal consequences that in the normal environment are not of selective significance but that in some odd environment may prove useful and thus be selected. Note that this is how virtually all major and all or most minor adaptations arise. Thus arose flight, sight, hearing, lungs, jaws, you name it.

Here is my puzzle. Do you think you could list ahead of time all the possible Darwinian preadaptations open to, say, current life-forms? More formally, can you finitely prestate all possible preadaptations? I have yet to meet someone who thinks the answer is yes. It's not clear how even to start to list the set of such preadaptations. Now, I deeply believe this statement to be true, but I do not know whether it is an empirical claim or a mathematical one, nor do I, at present, know how to prove it.

And herewith the story of Gertrude the squirrel. She lived 23.25 million years ago and had a single Mendelian dominant mutation that left her with two folds of skin, each stretching from wrist to ankle. Now, Gertrude was held to be so ugly that no one in her clan would talk to her, eat with her, or play with her. One day Gertrude was brooding in a magnolia tree when Bertha, an owl, in a neighboring pine, spotted her and thought, *Lunch!* Downward flashed Bertha, claws extended. Gertrude was terrified and, spreading her arms, jumped from the tree. *"Gaaaaahh!"* cried Gertrude. And Gertrude flew! And she got away from the befuddled Bertha. Thus Gertrude became a heroine in her clan and was married in a civil ceremony to a handsome squirrel a month later. Well, given that the mutation was dominant, soon there were lots of baby squirrels with the same two folds of skin, and that, dear reader, is why there are flying squirrels, more or less.

Now, could you have said, ahead of time, that the folds would work that day as wings? Maybe. Could you say that an obscure molecular mutation in a bacterium might allow the bacterium to detect a calcium current from a ciliate and take evasive action? I think not. More generally, I think we just don't have the concepts ahead of time to state what all possible Darwinian preadaptations might be, nor can we state what all possible environments might be.

Now note that this unpredictability is not slowing down evolution for a moment. Darwinian preadaptations are occurring all the time. So the biosphere appears to be doing something that we cannot describe beforehand—not because of quantum indeterminacy or chaotic dynamic behavior but because we don't have the concepts ahead of time.

That, in turn, means that the space of relevant possibilities of the biosphere—its phase space—cannot be prestated. Thus the biosphere is creative in a way we cannot prestate. And that stands in marked contrast to what Newton brilliantly showed us how to do: In physics, in general, one can prestate the set of all possibilities—that is, the phase space—then consult the laws and the initial and boundary conditions and calculate the forward trajectory of the particle in its phase space.

I suspect we cannot state the phase space, the space of possibilities, in the biosphere. You might, if you are a physicist, say, "Well, if you treat the system classically, there is always the classical n-dimensional phase space of all the positions and velocities of the particles in the [somehow isolated] system." That may be true, but then you do not yet know how to pick out the relevant collective variables (the wings of Gertrude) as the variables that will matter to the unfolding of the biosphere. So we seem to confront a limitation on knowledge that we had not recognized before. The evolving biosphere is doing something that cannot be foretold; we do not have the categories. The same, I think, applies to technological evolution: No one foresaw the Internet a century ago.

Interestingly, the fact that we cannot prestate technological possibilities, if true, cuts the core out of the contemporary reigning theory in economics: "competitive general equilibrium," which begins with the assumption that one can prestate all possible goods and services, then proves that markets clear—that is, all goods are sold to buyers at the contracted price. But we cannot state ahead of time all possible goods and services, so the reigning theory is wrong at the outset. And in focusing on market clearing, the reigning theory ignores a dominant fact about the global economy. The diversity of goods and services, like the diversity of species, has exploded. Why?

The fundamental laws of physics are time-reversible. Time enters, we have been led to believe, due to the irreversibility of the second law of thermodynamics. I do not disagree, but I wonder at the arrow of time implicit in the distinction between past and future in the biosphere due to Darwinian preadaptations. That appears to be not the second law at work but Darwin's variation and selection.

There is a final point. The universe is grossly nonrepeating, or nonergodic. Consider the number of proteins 200 amino acids long. Since there are 20 amino acids, the number of possible proteins 200 amino acids long is 20 raised to the 200th power, or roughly 10 raised to the 260th power. Now the number of particles in the known universe is about 10^{80}. Ask yourself whether all possible proteins could have been constructed since the Big Bang some 13 billion years ago. A fast chemical reaction takes a femtosecond (10^{-15} seconds). Thus the number of possible pairwise particle collisions on a femtosecond timescale since the Big Bang is 10^{80} x 10^{80} x 10^{33}, which equals 10 raised to the 193rd power. This is a big number, but tiny compared to the number of proteins of length 200, which is 10^{260}. Thus if the universe did nothing but make proteins, it would take 10^{67} repetitions of the life of the universe to make all possible proteins of length 200. And therefore, once above the level of atomic nuclei to the level of complexity of organic molecules (to say nothing of organisms, legal systems, or Chevy trucks), the universe is vastly nonrepeating. This means that when Gertrude pulled off her maiden flight she changed the physical and molecular unfolding of the universe. And so, too, for many or most of the actions of autonomous agents. We are travelers on a unique trajectory, and we have real physical consequences. Think not? We raced the Soviets to the moon, left some mass on the moon, and thereby changed the orbital dynamics of the solar system.

We've come a long way from my start—wondering what kind of system can act on its own behalf. While I am not sure I have stumbled on an adequate definition of life, I tend to think I may well have. Further, I'm sure that the next fifty years will see the construction of such systems experimentally. As they coevolve with one another, we will look on in amazement and not be able to say ahead of time what will happen. As the "chaotician" in *Jurassic Park* famously commented: "Life finds a way." He didn't add that we generally don't have much of a clue ahead of time what that way will be. Life is inherently open, and its understanding will require raising physics and chemistry to new levels, wherein the future is open rather than predictable in prestated categories.

❑

STUART KAUFFMAN, an emeritus professor of biochemistry at the University of Pennsylvania, is a theoretical biologist who studies the origin of life and the origins of molecular organization. He is a MacArthur Fellow and an external professor at the Santa Fe Institute. Twenty-five years ago, he developed the Kauffman models, which are random networks exhibiting a kind of self-organization that he terms "order for free." Dr. Kauffman is the founding general partner and chief scientific officer of The Bios Group, a company that applies the science of complexity to business management problems. He is the author of *Origins of Order* and *Investigations* and the coauthor (with George Johnson) of *At Home in the Universe: The Search for the Laws of Self-Organization.*

PART TWO

❑

THE FUTURE,
IN PRACTICE

RICHARD DAWKINS

❑

Son of Moore's Law

GREAT ACHIEVERS WHO HAVE gone far sometimes amuse themselves by then going *too* far. Peter Medawar knew what he was doing when he wrote, in his review of *The Double Helix*, "It is simply not worth arguing with anyone so obtuse as not to realize that this complex of discoveries [molecular genetics] is the greatest achievement of science in the twentieth century." Medawar, like the author of the book he was reviewing, could justify his arrogance in spades, but you don't have to be obtuse to dissent from his opinion. What about that earlier Anglo-American complex of discoveries known as the Neo-Darwinian Modern Synthesis? Physicists could make a good case for relativity or quantum mechanics, and cosmologists for the expanding universe. The "greatest" anything is ultimately undecidable, but the molecular genetic revolution was undeniably one of the greatest achievements of science in the twentieth century—and that means of the human species, ever. Where shall we take it—or where will it take us—in the next fifty years? By mid-century, history may judge Medawar to have been closer to the truth than his contemporaries—or even he—allowed.

If asked to summarize molecular genetics in a word, I would choose "digital." Of course, Mendel's genetics was digital in being particulate with respect to the independent assortment of genes through pedigrees. But the interior of genes was unknown and they could still have been substances with continuously varying qualities, strengths, and flavors, inextricably intertwined with their phenotypic effects. Watson/Crick genetics is digital through and through, digital to its very backbone, the double helix itself. A genome's size can be measured in gigabases with exactly the same precision as a hard drive is sized up in gigabytes. Indeed, the two units are interconvertible by constant multiplication. Genetics today is pure information technology. This, precisely, is why an antifreeze gene can be copied from an arctic fish and pasted into a tomato.

The explosion sparked by Watson and Crick grew exponentially, as a good explosion should, during the half century since their famous joint publication. I think I mean that literally, and I'll support it by analogy with a better-known explosion, this time from information technology as conventionally understood. Moore's Law states that computer power doubles every eighteen months. It is an empirical law without an agreed theoretical underpinning, though Nathan Myhrvold offers a wittily self-referential candidate: "Nathan's Law" states that software grows faster than Moore's Law, and that is why we have Moore's Law. Whatever the underlying reason, or complex of reasons, Moore's Law has held true for nearly fifty years. Many analysts expect it to continue for as long again, with stunning effects upon human affairs—but that is not my concern in this essay.

Instead, is there something equivalent to Moore's Law for DNA information technology? The best measure would surely be an economic one, for money is a good composite index of man-hours and equipment costs. As the decades go by, what is the benchmark number of DNA kilobases that can be sequenced for a standard quantity of money? Does it increase exponentially, and if so what is its doubling time? Notice, by the way (it is another aspect of DNA science's being a branch of information technology) that it makes no difference which animal or plant provides the DNA. The sequencing techniques and the costs in any one decade are much the same. Indeed, unless you read the text message itself, it is impossible to tell whether DNA comes from a man, a mushroom, or a microbe.

Having chosen my economic benchmark, I didn't know how to measure the costs in practice. Fortunately I had the good sense to ask my colleague Jonathan Hodgkin, professor of genetics at Oxford University. I was delighted to discover that he had recently done the very thing while preparing a lecture for his old school, and he kindly sent me the following estimates of the cost, in pounds sterling, per base pair (that is, per "letter" of the DNA code) sequenced. In 1965, it cost about £1,000 per letter to sequence 5S ribosomal RNA from bacteria (not DNA, but RNA costs are similar). In 1975, to sequence DNA from the virus .X174 cost about £10 per letter. Hodgkin couldn't find a good example for 1985, but in 1995 it cost £1 per letter to sequence the DNA of *Caenorhabditis elegans*, the tiny nematode worm of which molecular biologists are so rightly enamored that they call it "the" nematode, or even "the" worm. By the time the Human Genome Project culminated around 2000, sequencing costs were about £0.1 per letter. To show the positive trend of growth, I inverted these figures to "bangs for the buck"—that is, quantity of DNA that can be sequenced for a fixed amount of money, and I chose £1,000, correcting for inflation. I have plotted the resulting kilobases per £1,000 on a logarithmic scale, which is convenient because exponential growth shows up as a straight line. (See graph on following page.)

I must emphasize, as Professor Hodgkin did to me, that the four data points are back-of-the-envelope calculations. Nevertheless, they do fall convincingly close to a straight line, suggesting that the increase in our DNA sequencing power is exponential. The doubling time (or cost-halving time) is twenty-seven months, which may be compared with the eighteen months of Moore's Law. To the extent that DNA sequencing work depends upon computer power (quite a large extent), the new law we have discovered probably owes a great deal to Moore's Law itself, which justifies my facetious label, "Son of Moore's Law."

It is by no means to be expected that technological progress should advance in this exponential way. I haven't plotted the figures out, but I'd be surprised if, say, speed of aircraft, fuel economy of cars, or height of skyscrapers were found to advance exponentially. Rather than double and double again in a fixed time, I suspect that they advance by something closer to arithmetic addition. Indeed, the late Christopher Evans, as long ago as 1979, when Moore's Law had scarcely begun, wrote:

> Today's car differs from those of the immediate post-war years on a number of counts. . . . But suppose for a moment that the automobile industry had developed at the same rate as computers and over the same period: how much cheaper and more efficient would the current models be? . . . Today you would be able to buy a Rolls-Royce for £1.35, it would do three million miles to the gallon, and it would deliver enough power to

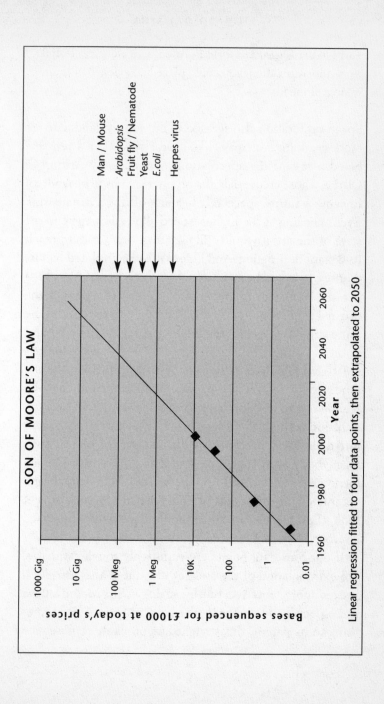

SON OF MOORE'S LAW

Man / Mouse
Arabidopsis
Fruit fly / Nematode
Yeast
E.coli
Herpes virus

Bases sequenced for £1000 at today's prices

1000 Gig
10 Gig
100 Meg
1 Meg
10K
100
1
0.01

1960   1980   2000   2020   2040   2060

Year

Linear regression fitted to four data points, then extrapolated to 2050

drive the *Queen Elizabeth II*. And if you were interested
in miniaturization, you could place half a dozen of them
on a pinhead.

Space exploration also seemed to me a likely candidate for
modest additive increase like motor cars. Then I remembered a fascinating speculation mentioned by Arthur C.
Clarke, whose credentials as a prophet are not to be ignored.
Imagine a future spacecraft heading off for a distant star.
Even traveling at the highest speed allowed by the current
state of the art, it would still take many centuries to reach
its distant destination. And before it had completed half its
journey, it would be overtaken by a faster vessel, the product of a later century's technology. So, it might be said, the
original ship should never have bothered to set out. By the
same argument, even the second spaceship should not
bother to set out, because its crew is fated to wave to their
great-grandchildren as they zoom by in a third. And so on.
One way to resolve the paradox is to point out that the
technology to develop later spaceships would not become
available without the research and development that went
into their slower predecessors. I would give the same answer
to anybody who suggested that since the entire Human
Genome Project could now be started from scratch and
completed in a fraction of the years the actual project
took, the original enterprise should have been postponed
appropriately.

If our four data points are admittedly rough estimates,
the extrapolation of the straight line out to the year 2050
is even more tentative. But by analogy with Moore's Law,
and especially if Son of Moore's Law really does owe something to its parent, this straight line probably represents a

defensible prognostication. Let's at least follow to see where it will take us. It suggests that in the year 2050 we shall be able to sequence a complete individual human genome for £100 at today's values (about $160). Instead of "the" human genome project, each individual will be able to afford his or her own personal genome project. Population geneticists will have the ultimate data on human diversity. It will be possible to work out trees of cousinship linking any person in the world to any other person. It is a historian's wildest dream. They will use the geographic distribution of genes to reconstruct the great migrations and invasions of the centuries, track voyages of Viking long ships, follow the American tribes by their genes down from Alaska to Tierra del Fuego and the Saxons across Britain, document the diaspora of the Jews, even identify the modern descendants of pillaging warlords like Genghis Khan.

Today, a chest X ray will tell you whether you have lung cancer or tuberculosis. In 2050, for the price of a chest X ray, you will be able to know the full text of every one of your genes. The doctor will hand you not the prescription recommended for an average person with your complaint but the prescription that precisely suits your genome. That is no doubt good, but your personal printout will also predict, with alarming precision, your natural end. Shall we want such knowledge? Even if we want it ourselves, shall we want our DNA printout to be read by insurance actuaries, paternity lawyers, governments? Even in a benign democracy, not everybody is happy with such a prospect. How some future Hitler might abuse this knowledge needs thinking about.

Weighty as such concerns may be, they are again not mine in this essay. I retreat to my ivory tower and more academic preoccupations. If £100 becomes the price of sequencing a human genome, the same money will buy the genome of any other mammal; all are about the same size, in the gigabase order of magnitude, as is true of all vertebrates. Even if we assume that Son of Moore's Law will flatten off before 2050, as many people believe Moore's Law will, we can still safely predict that it will become economically feasible to sequence the genomes of hundreds of typical vertebrate species per year, thousands of species of insects or other invertebrates, hundreds of thousands of bacteria, millions of viruses, and a disconcertingly variable number of amphibians and flowering plants. Having such a welter of information is one thing. What can we do with it? How shall we digest it, sift it, collate it, use it?

One relatively modest goal will be total and final knowledge of the phylogenetic tree. For there is, after all, one true tree of life, the unique pattern of evolutionary branching that actually happened. It exists. It is in principle knowable. We don't know it all yet. By 2050 we should—or if we do not, we shall have been defeated only at the terminal twigs, by the sheer number of species (a number that, as my colleague Robert May points out, is unknown to the nearest one or even two orders of magnitude).

My research assistant Yan Wong suggests that naturalists and ecologists in 2050 will carry a small field taxonomy kit, which will obviate the need to send specimens off to a museum expert for identification. A fine probe, hooked up to a portable computer, will be inserted into a tree, or a freshly trapped vole or grasshopper. Within minutes, the computer will chew over a few key segments of DNA, then spit out the species name and any other details that may be in its stored database.

Already, DNA taxonomy has turned up some sharp surprises. My traditional zoologist's mind protests almost undendurably at being asked to believe that hippos are more closely related to whales than they are to pigs. This is still controversial. It will be settled, one way or the other, along with countless other such disputes, by 2050. It will be settled because the Hippo Genome Project, the Pig Genome Project, and the Whale (if our Japanese friends haven't eaten them all by then) Genome Project will have been completed. Actually, it will not be necessary to sequence entire genomes to dissolve taxonomic uncertainty once and for all.

A spin-off benefit, which will perhaps have its greatest impact in the United States, is that full knowledge of the tree of life will make it even harder to doubt the fact of evolution. Fossils will become by comparison irrelevant to the argument, as hundreds of separate genes, in as many surviving species as we can bear to sequence, are found to corroborate each other's accounts of the one true tree of life.

It has been said often enough to become a platitude but I had better say it again: To know the genome of an animal is not the same as to understand that animal. Following Sydney Brenner (the single individual regarding whom, more than any other, I have heard people wonder at the absence so far of a Nobel Prize), I shall think in terms of three steps, of increasing difficulty, in "computing" an animal from its genome. Step 1 was hard but has now been completely solved. It is to compute the amino acid sequence of a protein from the nucleotide sequence of a gene. Step 2 is to compute the three-dimensional folding pattern of a protein from its one-dimensional sequence of amino acids. Physicists believe that in principle this can be done, but it is hard, and it may often be quicker to make the protein and see what happens. Step 3 is to compute the developing embryo from its genes and their interaction with their environment—which mostly consists of other genes. This is by far the hardest step, but the science of embryology (especially of the workings of Hox and similar genes) is advancing at such a rate that by 2050 it will probably be solved. In other words, I conjecture that an embryologist of 2050 will feed the genome of an unknown animal into a computer, and the computer will simulate an embryology that will culminate in a full rendering of the adult animal. This will not be a particularly useful accomplishment in itself, since a uterus or egg will always be a cheaper computer than an electronic one. But it will be a way of signifying the completeness of our understanding. And particular implementations of the technology will be useful. For instance, detectives finding a bloodstain may be able to issue a computer image of the face of a suspect—or rather, since genes don't mature with age, a series of faces from babyhood to dotage!

I also think that by 2050 my dream of The Genetic Book of the Dead will become a reality. Darwinian reasoning shows that the genes of a species must constitute a kind of description of the ancestral environments through which those genes have survived. The gene pool of a species is the clay which is shaped by natural selection. As I put it in *Unweaving the Rainbow*:

> Like sandbluffs carved into fantastic shapes by the desert winds, like rocks shaped by ocean waves, camel DNA has been sculpted by survival in ancient deserts, and even more ancient seas, to yield modern camels. Camel DNA speaks—if only we could read the language—of

the changing worlds of camel ancestors. If only we could read the language, the DNA of tuna and starfish would have "sea" written into the text. The DNA of moles and earthworms would spell "underground."

I believe that by 2050 we shall be able to read the language. We shall feed the genome of an unknown animal into a computer that will reconstruct not only the form of the animal but the detailed world in which its ancestors (who were naturally selected to produce it) lived, including their predators or prey, parasites or hosts, nesting sites, and even hopes and fears.

What about more direct reconstructions of ancestors, Jurassic Park style? DNA in amber is, unfortunately, unlikely to be preserved intact, and no sons or even grandsons of Moore's Law are going to bring it back. But there probably are ways, many of them as yet scarcely dreamed of, by which we can use the copious data banks of modern DNA that we shall have even before 2050. The Chimpanzee Genome Project is already under way and, thanks to Son of Moore's Law, should be completed in a fraction of the time taken by the human genome.

In a throwaway remark at the end of his own piece of millennial crystal-gazing ("Theoretical Biology in the Third Millennium," 1999, *Philosophical Transactions of the Royal Society, B*), Sydney Brenner made the following startling suggestion. When the chimpanzee genome is fully known, it should become possible, by a sophisticated and biologically intelligent comparison with the human genome (the two differ in only about 1 percent of their DNA letters), to reconstruct the genome of the ancestor we share. This animal, the so-called "missing link," lived between 5 million and 8 million years ago, in Africa. Once Brenner's leap is accepted, it is tempting to extend the reasoning all over the place, and I am not one to resist such temptation. The Missing Link Genome Project (MLGP) completed, the next step might be to line up the MLG with the human genome for a base-by-base comparison. Splitting the difference between the two (in the same kind of embryologically informed way as before) should yield a generalized approximation to *Australopithecus*, the genus of which Lucy has become the iconic representative. By the time the LGP (Lucy Genome Project) has been completed, embryology should have advanced to the point where the reconstructed genome could be inserted into a human egg and implanted in a woman, and a new Lucy born into the light of today. This will doubtless raise ethical worries.

Though concerned for the happiness of the individual australopithecine reconstructed (this is at least a coherent ethical issue, unlike fatuous worries about "playing God"), I can see positive ethical benefits, as well as scientific ones, emerging from the experiment. At present we get away with our flagrant speciesism because the evolutionary intermediates between us and chimpanzees are all extinct. In my contribution to The Great Ape Project, initiated by the distinguished Australian moral philosopher Peter Singer, I pointed out that the accidental contingency of such extinction should be enough to destroy absolutist valuings of human life above all other life. "Pro life," for example, in debates on abortion or stem cell research, always means pro human life, for no sensibly articulated reason. The existence of a living, breathing Lucy in our midst would change, forever, our complacent human-centered view of morals and politics. Should Lucy pass for human? The absurdity of the question should be self-evident, as in those South

African courts trying to decide whether particular individuals should "pass for white." The reconstruction of a Lucy would be ethically vindicated by bringing such absurdity out into the open.

While the ethicists, moralists, and theologians (I fear there still will be theologians in 2050) are busy agonizing over Project Lucy, biologists could, with relative impunity, be cutting their teeth on something even more ambitious: Project Dinosaur. And they might do it by, among other things, helping birds to cut teeth as they haven't done for 60 million years.

Modern birds are descended from dinosaurs (or at least from ancestors we would now happily call dinosaurs if only they had gone extinct as dinosaurs should). A sophisticated "evo-devo" (evolution and development) interpretation of modern bird genomes and the genomes of other surviving archosaurian reptiles such as crocodiles might enable us, by 2050, to reconstruct the genome of a generalized dinosaur. It is encouraging already that a chicken beak can be experimentally induced to grow tooth buds (and snakes induced to grow legs), indicating that ancient genetic skills still linger. If the Dinosaur Genome Project is successful, we could perhaps implant the genome in an ostrich egg to hatch a living, breathing, terrible lizard. Jurassic Park notwithstanding, my only anxiety is that I am unlikely to live long enough to see it. Or to extend my short arm to a new Lucy's long one and shake her tearfully by the hand.

❏

RICHARD DAWKINS, an evolutionary biologist, is the Charles Simonyi Professor for the Public Understanding of Science at Oxford University. He is the author of *The Selfish Gene,*

The Extended Phenotype, The Blind Watchmaker, River out of Eden, Climbing Mount Improbable, and *Unweaving the Rainbow.* He is a Fellow of the Royal Society and also a Fellow of the Royal Society of Literature, holds honorary doctorates in literature as well as in science, and is one of the few living scientists to have made it into the Oxford Dictionary of Quotations. He won the International Cosmos Prize for 1997, and is the 2001 winner of the Kistler Prize.

PAUL DAVIES

❑

Was There a Second Genesis?

"THAT MARS IS INHABITED by beings of one sort or other is as certain as it is uncertain what those beings may be." With these dramatic words, the American astronomer Percival Lowell informed the world about a network of canals he thought existed on the Red Planet. Lowell conjectured that Mars was a dying, drying planet, whose inhabitants built channels to bring meltwater from the polar caps to the arid equatorial regions. He produced elaborate maps to support his theory.

That was in 1906, when the idea of life on Mars seemed entirely plausible. H. G. Wells brilliantly exploited this belief in his highly successful book *The War of the Worlds*, written in 1898. Many astronomers at least paid lip service to the possibility that Mars might harbor some form of life. Then in the 1960s the Mariner space probes sent to Mars failed to reveal any sign of the much discussed canals. In 1976 two NASA spacecraft landed on Mars and found a desolate, lifeless terrain. They scooped up dirt and analyzed it for microbes and traces of organic compounds. Nothing was found. The Red Planet seemed to be a freeze-dried desert bathed in lethal ultraviolet rays. In short, Mars looked dead—very dead.

Recently, however, opinion has begun to shift. Maybe we were too hasty in writing off Mars as an abode for life. The early photographs of the surface from the Mariner series showed dried-up river channels, while the much more detailed pictures from the Mars Global Surveyor orbiter, obtained in the last few years, reveal what look like flood-plains, dried-up lake beds, and even the hint of an ancient ocean. Evidently Mars was once warm and wet, and not unlike our own planet. Could life have flourished there in the remote past? Might it still be clinging on today in some obscure niche?

There is a good chance we shall learn the answers to these questions in the next fifty years. The nascent subject of astrobiology looks set to advance dramatically over the coming decades, and research projects such as NASA's Origins Program promise to make the technology available for us to seek out life beyond Earth and address the age-old question: Are we alone? As the one planet (besides Earth) in our solar system accessible to human exploration, Mars will receive special attention. Motivation to go there is strong. It could be our only chance of studying a second genesis, another location in the universe where life has emerged from nonlife.

Where exactly must we look for life on Mars? Its surface is forbiddingly hostile to any form of familiar life, for which liquid water is essential. There is abundant water ice at the poles, but the temperature is far too low for it to melt. Even if it did, the liquid would rapidly evaporate, because the atmospheric pressure on Mars is less than 1 percent of that on Earth. In the past, Mars must have had a much thicker atmosphere, laden with greenhouse gases such as carbon dioxide; this would have elevated the temperature and provided sufficient pressure for liquid water to remain on the surface for long periods. Estimates suggest that this "Garden of Eden" phase ended as much as 3.5 billion years ago, although sporadic warming may have occurred since. Three and a half billion years is a long time for life to remain dormant on the Martian surface, so the best hope for finding life there today is in the subsurface zone. Over the past twenty years, scientists have been astonished to find microbes dwelling deep inside Earth's crust. Organisms have also been found far beneath the seabed. The depth of this hidden biosphere extends in places to several kilometers. Because the temperature rises with depth, deep-living organisms tend to be heat-loving organisms, known as thermophiles; in some cases, they thrive at temperatures above the normal boiling point of water. The energy source for many species of subsurface life derives not from sunlight but from chemical and thermal energy. Some microbes can take gases and minerals percolating up from Earth's crust and turn them directly into biomass, thereby supporting an entire food chain independent of surface life.

The discovery of subsurface life capable of sustaining itself without sunlight has greatly boosted hopes for life on Mars. Like Earth, the Red Planet has a hot interior, evidenced by its extensive volcanoes, some of which may still be active. There are undoubtedly hot spots underground on Mars, where volcanic heat has melted the permafrost, providing liquid reservoirs that could host primitive life. Subsurface Martian life might reveal its presence through exuded gases such as methane seeping to the surface. Future space probes will be able to search for tiny concentrations of such biogenic gases in the Martian atmosphere. To properly study any underground Martian microbes, it will be necessary to probe deep beneath the surface. Nobody knows just how deep; estimates vary from a few meters to kilometers. Planned missions, such as the European Space Agency's aptly named *Beagle 2*, set to launch in June 2003, will incorporate penetrators and drills; however, they are unlikely to go deep enough to reach any extant organisms.

Rock samples from the surface, returned to Earth for analysis, could give clues about past life on Mars. The best evidence would be the discovery of microfossils. In 1996, NASA scientists announced that an Antarctic meteorite, blasted off Mars by asteroid impact at least 16 million years ago, had been found to contain tiny features reminiscent of fossilized microbes. The meteorite, known as ALH 84001, has been subjected to intense scrutiny, but though the jury is still out, the prevailing view is that ALH 84001 will not provide definitive evidence of Martian life.

The first Mars sample-return mission is being planned by NASA for the end of this decade. A robotic probe will attempt to collect a selection of interesting-looking rocks from the Martian surface to be sent back to Earth for analysis. To be absolutely safe, the rocks will be placed in strict quarantine, not only to keep them from being contaminated. by Earthly biota but as a precaution against the release of any virulent Martian bacteria. But the chances that a killer plague from Mars will wipe us all out are extremely remote. An average of one Martian meteorite per month hits Earth; over our geological history, billions of tons of Mars rocks have come here. Given this traffic in material, it seems likely that ancestors of any Martian microbes present in the returned samples would have hitched a ride here and infected us already. Experiments with artillery shells and centrifuges have shown that microbes can easily withstand the shock of ejection from

Mars. Once in space, the cold vacuum conditions act as a preservative. Some bacteria form tough spores when stressed and can survive for an enormous period of time in a dormant state. The greatest hazard in traveling through interplanetary space is radiation, but microbes cocooned inside a rock a couple of meters across would be shielded from solar ultraviolet, solar flares, and all but the most energetic cosmic rays. Calculations suggest that hardy bacteria could endure millions of years in orbit around the sun if ensconced in a suitable rock. The final hazard—high-speed entry into Earth's atmosphere—need not present a problem, because the frictional heat would not have time to penetrate to the meteorite's interior. All in all, there seems to be no major impediment to the successful transfer of viable Martian organisms to Earth.

Follow-up sample-return missions are already in the planning stage; more careful consideration will be given to the landing site once the robotic technology has been proved. The use of rover vehicles is likely to significantly improve with the employment of onboard supercomputers, neural nets, and sophisticated sensor technology. Instead of the lumbering, cumbersome vehicles such as the Sojourner rover on NASA's 1997 Pathfinder mission, smart rovers will be able to explore on their own, making the necessary on-the-spot decisions about terrain, rock samples, and so on with no help from mission control. The development of a Mars plane that can swoop and glide over the barren surface will vastly improve the surveying techniques at present limited to orbiters.

These technological advances should enable the Martian surface material to be studied in great detail. Nevertheless, it won't be easy to find Mars rocks containing 3.5-billion-year-old fossils (let alone live microbes). On Earth there are only a handful of places where fossils of this age are found, and only then after very careful selection. The chances of an unmanned probe collecting a Mars rock with fossils in it are slim. It may be that after some decades of intensive investigation, the issue of whether or not there is, or was, life on Mars will remain unresolved. If so, the final hope for settling the issue will rest with a crewed expedition.

Getting people to Mars won't be cheap. The price tag for sending four astronauts will be at least tens of billions of dollars. However, some imaginative cost-cutting is possible. The independent engineering consultant Robert Zubrin has pointed out in his 1996 book *The Case for Mars* that a major part of the expense in a Mars expedition comes from transporting the fuel for the return journey. But this may not be necessary. Mars has both water and carbon dioxide, which together can be turned into methane, a good propellant. Zubrin envisages sending a chemical reactor on ahead to the Martian surface and waiting for it to generate a full tank of fuel before launching the expedition. The astronauts would face many hazards on the interplanetary journey, which would last several months in each direction; however, once they had landed and set up a base, life need not be too dangerous. As Zubrin has remarked, the surface of Mars is the second safest place in the solar system.

In the Zubrin scheme, the astronauts remain on the surface for two years, during which time they carry out wide-ranging exploration in rover vehicles and search for signs of life. Drilling equipment could be dispatched ahead of the crew, enabling rock samples from the deep subsurface to be extracted. Ideally the first expedition would return to Earth when a replacement contingent arrived, thus establishing a continuing human presence on the planet.

Mounting a Mars expedition would be much more ambitious than the Apollo moon landings, and many technological problems need to be addressed first. For example, the medical problems associated with prolonged weightlessness could prove serious; the use of the international space station should provide valuable experience. Although it may take several decades to plan, I see little reason why men and women should not go to Mars by 2050.

What if we do find life on Mars? The significance of the discovery will hinge crucially on whether Martian life is the same as terrestrial life. This is important, given the possibility that Mars and Earth have cross-contaminated each other. There is substantial traffic of material not only from Mars to Earth but also (since Earth occasionally suffers large asteroid and comet impacts, too) in the opposite direction—although fewer rocks go the other way, because of Earth's deeper gravity well. Microbes may have been transported in either or both directions by this process. This intermingling of the two biospheres would considerably complicate the picture. It is entirely possible that life started on one planet and spread to the other before a second genesis could happen. It is a moot point whether introduced life would rapidly commandeer all available niches and food resources, thus stifling a second genesis, or whether two different biological systems could coexist on the same planet.

Mars seems the more favorable planet for life to have begun. Being smaller than Earth, it cooled quicker, and it may have been ready for life as long as 4.4 billion years ago. By contrast, Earth may not have been habitable until about 3.9 billion years ago. Both Mars and Earth were subjected to a ferocious bombardment by giant asteroids and comets for at least 700 million years after the formation of the solar system 4.5 billion years ago. The biggest impact events would have been so violent that they would have sterilized the entire planet, swathing it in incandescent rock vapor at a temperature of 3,000°C. This global furnace would have sent a heat pulse a kilometer into the ground, killing anything that wasn't well below the surface—but no organisms would have settled at very great depths because it would be too hot to live there. So there would have been a comfort zone, bounded below by the internal heat of the planet and above by the heat pulses from major impacts. On Mars, this comfort zone for deep-living microbes would have been deeper sooner, making it a better bet for life to have established itself early.

All life on Earth is interrelated—that is, it descended from a common ancestor. The multifarious species inhabiting our biosphere are just different branches on a universal tree of life. If life started on Mars and spread to Earth, then any extant or relic Mars life would represent just another branch on this tree—maybe a lower, older branch but one sharing a common origin with terrestrial life. By 2050, gene sequencing techniques will be highly automated and the equipment transportable. It should be possible to carry out the necessary analysis at the Martian base, removing the need for quarantining procedures.

If Martian life turns out to be the same as terrestrial life, then Mars will have failed to provide the second sample of life we urgently seek. It would still be possible to assert that the origin of life was a freak accident, unique in the universe. To settle the matter of life's uniqueness or ubiquity, we would need to look farther afield. The only other body in the solar system suspected of having substantial quantities of liquid water is Europa, a moon of Jupiter. It has an icy crust beneath which probably lies a liquid ocean, warmed by tidal friction as Europa orbits the giant planet. Being so distant, Europa is unlikely to have been biologically contaminated by Earth or Mars. Unfortunately a crewed expedition to Europa is out of the question with any foreseeable technology, even in the next fifty years. However, an unmanned probe will likely be sent there in the next thirty years. The challenge will be to penetrate the thick ice layer. One way this might be done is to equip the probe with a small nuclear reactor so that it can melt its way down through the ice. It could then dispatch a small submarine to explore the dark ocean beneath.

Astrobiologists agree that there is scant chance of finding any type of extraterrestrial life in the solar system more advanced than simple bacteria. Complex life-forms probably need a planet very much like Earth, with a thick atmosphere, liquid water, an ozone layer, and plate tectonics to recycle atmospheric gases such as carbon dioxide. The search for Earth-like planets in other star systems will be a major research theme in the coming decades. The distances to the stars are so enormous that their exploration is highly unlikely even in fifty years. Without a revolution in propulsion technology, any craft sent beyond the solar system would take thousands of years to reach their destinations; the search for other Earths must for the foreseeable future depend on improved observational technology. Astronomers have discovered dozens of extrasolar planets in recent years using ground-based optical telescopes, but so far the techniques used are not sensitive enough to spot a planet the size of Earth in a similar orbit around another star. To achieve this will require a large ultraprecise space-based optical system capable of picking out the feeble reflected light of the planet from the intense glare of its parent star and then analyzing the spectrum for telltale signs of life, such as oxygen in the planet's atmosphere. This is a key objective of NASA's Origins Program.

One proposal is for a system of four optical telescopes flying in carefully synchronized formation to create a huge interferometer that can resolve distant astronomical objects in unprecedented detail. This system—called TPF, for Terrestrial Planet Finder—could be in solar orbit by 2016. If the Planet Finder proves successful, it will be followed by PI, the Planet Imager—an even larger interferometer with resolving power equivalent to that of a single telescope 360 kilometers wide! This would provide close-up pictures of whatever Earth-like extrasolar planets have been found, revealing the activities of any surface life. It is an arresting thought that within a hundred and fifty years of Lowell's painstaking but misguided observations of a Martian canal network, we could image such structures in another star system, many light-years away.

Of course, we would be extremely lucky to find complex, intelligent life in our close galactic neighborhood. Other star systems may well have planets like Earth on which life has remained stuck at the bacterial level. The existence of complex life on Earth probably depends on certain rather special features of our solar system. For example, our planet has an unusually large moon, which helps stabilize its motion and prevent large climatic variations. The moon probably formed from Earth's outer layer, as a result of a glancing blow by a Mars-size object during the formation of the solar system, a chance event unlikely to happen often. The planet Jupiter also plays a crucial role by sweeping up comets that might otherwise crash into Earth and cause frequent mass extinctions. These and other circumstances, such as the chemical composition of our planet and the stability of the sun, suggest that habitable planets like Earth are rather rare in the Galaxy.

The search for life beyond Earth lies poised at the threshold of success. A great deal hinges on the outcome, because the search for life elsewhere is also a search for ourselves—who we are and what our place might be in the great cosmic scheme. If life is a stupendous chemical fluke confined to our little corner of the universe, and intelligent beings like ourselves are unique, our responsible stewardship of planet Earth becomes all the more important. If we do find a second genesis, it will forever transform our science, religion, and worldview. A universe in which nature's laws are biofriendly is a universe in which life is a fundamental, rather than an incidental, feature. It is a universe in which we can truly feel at home.

❑

PAUL DAVIES, a theoretical physicist and a visiting professor at Imperial College London and the University of Queensland, is the author of such best-selling popular science titles as *About Time, The Mind of God*, and *The Fifth Miracle: The Search for the Origin of Life*. Davies's research has been mainly in the field of quantum gravity and cosmology; however, his interests are much wider, ranging from particle physics to astrobiology. He is currently working on the problem of biogenesis and the role of cosmic impacts on the early development of life. For years he has written and lectured about the deeper implications of science, for which work he was awarded the $1 million Templeton Prize in 1995.

JOHN H. HOLLAND

❑

What Is to Come and
How to Predict It

TO UNDERTAKE LONG-RANGE predictions is Promethean:
The fate of the predictions, if not of the predictor, is likely
to be unhappy. Still, the challenge is hard to ignore. In my
decision to tread where I shouldn't, the clincher was the
realization that there is an oblique course to the objective,
by concentrating on the underpinnings of the prediction
process itself and treating the predictions as a kind of illus-
tration.

The single most important factor in making an accurate prediction is the level of detail. Experienced players of
board games can often predict a win or a loss after the first
few moves, but they rarely attempt to predict details of the
ending configuration. On a more sophisticated level, many
biologists would make the general prediction that evolving life-forms will be a common feature of Earth-like planets, but few would make the specific prediction that such
an evolving ecosystem inevitably produces a primatelike
organism. A similar point holds for almost any attempt at
prediction.

The common way of making predictions is to examine

extensions of current trends. Using this technique, we predict everything from future income, be it gross domestic product or personal finance, to population changes, be they changes in species numbers or the depradations of disease. Such predictions can be valuable in the short term, but trends are fallible guides for longer periods, unless the underlying processes have great "inertia," as in the case of population growth or the buildup of greenhouse gases.

Fifty years is a long time on the current technological and social scale, even when we are looking at population growth and greenhouse effects. Moreover, on that timescale the prevalent features are heavily influenced by what are now called complex adaptive systems. Complex adaptive systems (CAS) consist of many interacting components, called agents, that adapt to (or learn from) each other as they interact. Stock markets and immune systems are familiar examples of CAS. Even on relatively short time-scales, CAS exhibit a range of nonadditive (nonlinear) effects: self-organization, chaos, fractal attractors, frozen accidents, lever points, and the like. As a result, the actions of the components cannot be summed to give an overall trend. Moreover, we have only bits and pieces of a theory of CAS. Because we lack an overarching theory, we have no general, principled way for determining the influence of these nonadditive effects. As a consequence, when CAS are involved, prediction is fraught with hazard.

Despite this warning, I do think it is possible to uncover some plausible alternative futures. This possibility turns on our ability to construct models. Computer-based models are at present our most powerful tool for examining alternate histories. The models let us see the results of different action sequences, in the manner of an expert pilot using a flight simulator to "test the envelope" of a new aircraft design. Generating predictions via models has several distinct advantages:

1) The assumptions underlying the predictions are made explicit, so others can judge the assumptions' relevance and the plausibility of the predictions. Indeed, others can use or modify the assumptions to make their own predictions, enriching the overall enterprise.
2) Well-constructed models are modular, so that errors can be traced to relevant modules and missing responses, suggesting revisions or new modules for improvement.
3) Models encourage replay under different conditions and actions (testing the envelope) in order to demonstrate the robustness of the predictions.

Though the advantages of modeling are important for any attempt at prediction, the approach has a distinct disadvantage for the purposes of this essay: Building computer-based models is a time-consuming task, taking anywhere from months to decades; witness the attempts to model climate change. The time allotted for preparing this essay does not allow a full-bore modeling approach. Still, one can accumulate crude descriptions of some of the building blocks that would be the initial steps in constructing such a model. Crude descriptions allow video-game-like images, and human intuition often works well with such images.

As in designing animations, the starting point in setting up a video-game-like model is to determine what changes slowly or not at all. Invariants and slowly changing quantities provide a framework that shores up the predictions. Once we construct this framework, we want to incorporate the elements that seem easier to control or predict. It is usually easier to predict technological change than social change, though some writers of fiction—Jules Verne, H. G. Wells, and Arthur C. Clarke come to mind—have made a good stab at long-term social prediction. *Paris in the Twentieth Century*, Verne's 1863 novel set in 1960 Paris, makes a remarkable number of "hits" despite the intervention of unprecedented political creeds, wars, and boundary changes. Nevertheless, successes in long-range social prognostication are rare. Gordon Moore's prediction of an eighteen-month doubling time for computer hardware capability has held for several decades, while I know of no one who anticipated such social phenomena as Amazon.com or e-Bay at the time Moore's prediction was made.

So I'll start with what appears to be the easier task, describing some of the technological changes I think likely in several interlocking areas: computerization and robotics, biology, transportation, and space exploration. Then I'll infer some of the social consequences of those changes: their effects on population, planning and education, privacy, medicine, and a new age of exploration.

A Technological Frame

The broad outline of the future of the computer and its offshoots, like the Internet, is relatively easy to foresee on a ten- to twenty-year scale. Moore's law for hardware advance will continue to hold, and the present snail's pace for doubling software proficiency—somewhere between one and two decades per doubling—also seems likely to persist. The slow pace of the latter is at least as important as the rapid pace of the former. Though we now have some ability to provide computers with simple, task-oriented learning, we are still little better off than we were in the middle of the last century when it comes to giving them broader human abilities, such as flexible pattern recognition (recognizing familiar objects in a cluttered natural setting) or language understanding (for example, understanding a novel). We still have only the most sketchy ideas for providing computers with capabilities for invention, reasoning by analogy and common sense, making hypotheses, and the like.

Moore's law does not help much when we confront CAS problems, because small changes in problem size can cause great increases in complexity. To take a simple example, the number of possible ten-move sequences in the game of Go can be increased by a factor of five simply by adding a single new row and column to the 19 x 19 game board, without changing the rules at all! If this is so for a game defined by half a dozen rules, what is the case for CAS, such as markets and governments, where even crude models may involve dozens of "laws" describing the interactions? Such problems will not yield to the hardware and hacking approach exemplified by Deep Blue's triumph over Garry Kasparov. Yet humans deal with CAS on a routine basis, often quite proficiently. To paraphrase the AI pioneer Marvin Minsky, the wonder is not that Deep Blue can play chess at the level of a world master but that humans, with so much less ability to search in detail at, say, a depth of ten moves, can challenge the computer at that level. Attempts to discover mechanisms that generate thought and consciousness have occupied humankind since the beginning of recorded history. Most psychologists now believe that consciousness is tied to the activity of neurons in the central nervous system, but we still know surprisingly little about the relation between consciousness and neural activity. Unraveling this relation has proved to be notoriously difficult, and I do not expect sudden "solutions" in the next fifty years.

A Social Frame

In making social predictions, we at least know that human nature has changed slowly, if at all, over the millennia. There were Roman senators who made substantial fortunes by cornering the grain market when they received early warning through government couriers that there would be a shortfall in the Carthaginian grain crop, Rome's "bread basket." Both the cupidity and the use to which such short-term predictions were put have changed little in the ensuing two thousand years.

In addition to the constant provided by human nature, there are problems produced by processes that have high inertia. Population size and population-influenced processes, such as the accumulation of greenhouse gases, provide prime examples. They have an inherently slow turnover, because of underlying generation times. Much of our social agenda is influenced by problems that are not subject to a quick fix. Fixes in this realm require plausible predictions of the long-term effects of current actions.

The present sustained, extensive effort at increasing our understanding of biological processes is likely to persist because of the huge associated social and financial benefits. However, here is a case where details of that effort—what we will exploit and what we won't—are subject to a wide range of political, economic, and personal considerations. The conflict between short-range benefit and long-range exploration is particularly strong in this area.

Technological Change

The continued reign of Moore's law makes it feasible to further miniaturize and combine several common late-twentieth-century devices. We will achieve a wristwatch-size worldwide communicator/videocam/computer/animator/global positioner/notepad, with a 3-D projection display (similar to the projector that R2D2 uses in *Star Wars*), allowing heads-up, hands-on control. This device will become as common as a wristwatch, too; and, with its video-game-like interfaces and user-oriented learning, it will finally be truly as easy to use as a notepad.

If we are to have computers that can deal autonomously with complex adaptive systems, we will need software that can routinely and effectively exercise human-level flexibility and intelligence. I think this can be accomplished only by software that can learn and evolve. Because problems involving CAS are important and pervasive—they range from such social phenomena as inner-city decay or fluctuations in global trade to environmental matters like invasions by exotic organisms or the sustainability of eco-systems—there will be steadily increasing emphasis on software with these capabilities. Even at the present slow pace of improvement, there will be increased use of software that modifies itself on the basis of experience to meet the idiosyncratic needs of individual users. In fifty years, we will probably have robots that can act as trained assistants, though they will be brittle in the face of the unexpected. I do not think we will have "conscious" robots within fifty years, though I do think it will happen eventually.

Our understanding of life and living organisms has been much enhanced by computers and automated laboratory equipment. By the mid-twenty-first century, much of medicine as it was practiced in the latter part of the twentieth century—for example, using surgery, chemotherapy, and radiation to treat cancer—will look as ineffective as the bloodletting of earlier centuries. We will likely be able to produce life in a test tube by starting with simple, nonliving biochemicals, with all that that implies for engineered solutions to diseases. And we will almost certainly be able to produce artificial immune systems that can counter both living viruses and computer viruses. The power of the immune system we're born with is often underestimated; it combats a wide range of subtle invaders so efficiently that most of us escape diseases for long spans of time. This is the more remarkable when we realize that these disease-free spans, in terms of body-cell generations, are comparable to the number of human generations between the Middle Ages and the present. The diagnostic capabilities of artificial immune systems will also help us solve some of the difficulties in relating sequenced genomes to the underlying complex networks of signaling molecules—the so-called biocircuits—that give biological cells their coherence and flexibility.

There is one technological area in which change is long overdue: ground transportation. The twentieth-century automobile gave tremendous mobility to citizens in the developed countries, with consequent freedom from serf-like bonds to place. But we are still confined to the congestion of Roman roads. Though this status quo is supported by a tremendous infrastructure of companies and lobbies, there are substantial changes in compact energy delivery on the horizon. Computer guidance and global positioning make possible flexible individual transport, without confinement to roads for routing. It's a fifty-fifty bet that there will be some form of quiet, triphibious (land, water, and air) individual transporter within fifty years. Though there will have to be approved rights-of-way for this new kind of transport, there will be large savings in infrastructure costs, such as highway and bridge maintenance.

Finally, I think we will recover from our long detour on the path to space exploration. Forty years ago, we were capable of *flying* to the edge of space with the X-series and the Blackbird. But we ditched all that—often destroying critical production facilities and even blueprints—to take up the well-engineered but dead-end approach of attaching pods to booster rockets. However, we are drawing near to the end of the detour for a variety of reasons:

1) We are once again beginning to investigate propulsion systems, such as SCRAM jets, that can fly us into space.
2) There are clear scientific, military, and economic advantages for nations that can maneuver freely in interplanetary space, much like the advantages that accrued to nations that could cross the open ocean in the fifteenth and sixteenth centuries.
3) Astronomy in the latter part of the twentieth century, with particular kudos to the Hubble Space Telescope, has shown us what wonders await us "out there."

Social Change

The number-one priority on a fifty-year scale is bringing Earth's human population down to a value more in line with renewable resources. Some of our most serious large-scale problems—inadequate food production, forest depletion, global warming, energy shortages—are traceable to a surplus of humans relative to resources. The sheer numbers of people rubbing shoulders produces physical and psychological stress not amenable to technological fixes. Whole nations now give this problem high priority—witness China—so I think that within fifty years population numbers will come under control without massive disasters like the Black Death or global war.

We are in fact becoming more sensitive to other such long-range problems, looking more carefully at options and alternatives. Traditionally, skills in exploring alternatives have been sharpened via board games, war games, and the like, but the context has always been limited. Video games have broadened that context—games like SimCity and Civilization substantially increase our sensitivity to intricate sociopolitical interactions—and the interfaces are much more realistic, allowing the ordinary citizen to explore options with ease. When this trend becomes more tightly linked to sophisticated simulations in everything from climate to artificial intelligence, there will be an order-of-magnitude increase in the number of people who routinely explore future options in a principled way. The miniaturized general-purpose device I mentioned earlier—call it a "planner"—will accelerate this process, tying "lookahead" into everyday situations. As with video games, programming expertise will not be required, only the intuitions and explorations suggested by experience. "Planners" will enable the post-video-game generation to examine the consequences of familiar actions, using realistic and controllable 3-D interfaces that have adapted to the capabilities of the user. In effect, we will have flight simulators for "testing the envelope" of social and political decisions.

Of course, there are social problems associated with these "planners." One is a problem that already exists: an increasing distance in knowledge and income between those who enter professions or disciplines where planners are a natural aid and those who don't want to or can't. In developed countries, almost everyone will use the planners, which will be the mid-twenty-first-century counterpart of the telephone, to probe their options; elsewhere the knowledge and income gap will widen precipitously. Currently less than 15 percent of the inhabitants of South America attend school to the ninth grade or beyond; not many of them have the regular use of telephones, and even fewer will have regular use of planners.

There is another, broader problem that will plague us all: holding on to privacy and the freedom to proceed without continual surveillance. The videocam/quick-communication capability of the planner makes everyone a newsperson. This has a good side: Rape, mugging, robbery, and other crimes that thrive on concealment and lack of evidence will become increasingly difficult when an unfolding situation can be broadcast instantly. On the other hand, and much more important, the potential for invading private affairs will make the current media predilection for reporting anything of "public interest" (disasters, human foibles, and the like) seem pale in comparison. Privacy and freedom from intrusion lie at the heart of an enlightened democracy ("A man's home is his castle"). When these rights are abrogated, tyranny quickly appears. By the mid-twenty-first century, it will be possible technologically to track the detailed movements of any individual. We will be as much under the thumb of outside powers as the serfs of the Middle Ages, who required permission to travel beyond the closest village. As with limits on freedom of expression (you can't yell "Fire!" in a crowded theater), the challenge here is to forge a mixture of law and custom that severely limits the rights of government and individuals to intrude. It is an open question whether or not we can succeed in this endeavor—shades of *1984*, one hundred years later.

Our enhanced understanding of biology will give us unprecedented control over disease and injury, and the freedom from pain that that implies. At the same time, we will have enhanced opportunities for waging biowarfare and for making mistakes in genetic engineering. But here I think the defense will keep pace with, or even lead, the offense. Artificial immune systems constitute powerful protection for both natural and artificial systems. The ability of artificial immune systems to discover biomolecules that counter unusual antigens, combined with the technological automation of drug design and production, will finally bring drug costs down, even in small markets (rare diseases), much as the inexpensive production of CDs made it possible to record music for very limited sets of listeners. This reduction in treatment costs, combined with the diagnostic capabilities of artificial immune systems, should at last reverse the ever increasing costs of medicine.

Finally, our coming ability to maneuver in interplanetary space will rival the exploration of the New World in producing an age of discovery and excitement. Within fifty years we will probably have bases on the moon, on Mars, and circling Jupiter. These bases will act much like the first European outposts in the New World in the fifteenth and sixteenth centuries, producing a steady stream of wonders that stimulate our imagination and curiosity. And being "out there" greatly increases our chances of receiving evidence (à la SETI) of other civilizations in our galaxy. Such an observation, if made, will have effects at least as great as the effects on medieval Europe of the rediscovery of the works of classical Greece.

❏

JOHN H. HOLLAND is a professor of psychology and of computer science and engineering at the University of Michigan at Ann Arbor, and an external professor and member of the board of trustees of the Santa Fe Institute. His main research interests are complex adaptive systems (natural and artificial), computer-based models of cognitive processes, and the construction of models for computer-based thought experiments. Known widely as the "father of genetic algorithms," he is a board member of the International Society for Genetic and Evolutionary Computation. His two most recent books are *Emergence: From Chaos to Order* and *Hidden Order: How Adaptation Builds Complexity*.

❑

The Merger of Flesh
and Machines

For at least the last five hundred years, science and technology have confronted us with generalizations that eroded our sense of ourselves and our world as unique and made us variously uncomfortable, enraged, and even violent. Early in the seventeenth century, with almost fifty years of careful observational data backing him up, Galileo clashed with the church over the position of the earth in the celestial scheme. Despite his tactical retreat in the face of religious intransigence, it soon became clear that the earth was not a unique object at the center of the universe but just one of several planets orbiting the sun. Later, of course, it was realized that the sun was just one of many stars, and still later that our galaxy was just one of many galaxies. Today we struggle intellectually over whether this humbling realization applies to our universe as well.

In his time, Charles Darwin generalized humans as just a part of the animal kingdom, directly related to it by bloodlines—a fact that even today is the subject of political bullying in the intellectual badlands of the United States. The twentieth century saw small embellishments of this idea when, following the work of Crick and Watson, it became clear that many of our most fundamental genes have diverged little from those of yeast or fruit flies. At the end of the century, we were faced with two more such generalizations: Perhaps our version of life did not originate here on Earth but was seeded from a life source on another planet. And finally we found that humans did not have as many genes as were expected, and in fact had fewer genes than many other animals, and even potatoes. We are not unique on that count, either.

Each of these generalizations has challenged our view of ourselves. We have become less special, part of a bigger reality. The loss of specialness has often been hard to bear, but slowly we have managed to adapt to the new worldview that each revelation has brought us. None of them has been abrupt. The discovery of extraterrestial intelligence, if it happens, might seem an abrupt intellectual jolt, and in some sense it would be; but even here we are becoming slowly sensitized, as more and more of us understand SETI, the search for extraterrestrial intelligence, well enough to volunteer our unused computer cycles for the effort. All these specieswide realizations have come upon us through many preparatory discoveries, arguments, and discussions. The climaxes may have been dramatic, but the signs were always there.

Now, at the beginning of the twenty-first century, we can see signs that the next fifty years promise another such generalization. Our very humanity will feel threatened, and that may well lead to violent wars over what are essentially intellectual and religious ideas. We already see the early skirmishes in these wars, and they are not at all pretty. The generalization we are facing is that we humans are machines—and as such, subject to the same technological manipulations we routinely apply to machines. To make matters slightly more complex, our technological infrastructure is going to change completely, just as it has over the last fifty years, and the technology of our bodies and of our manufacturing will be generalized as the same thing.

The central unstated tenet of modern molecular biology is that everything about living systems, ourselves included, is a product of molecular interactions. Modern biology is based on strict materialism. There is nothing else besides molecules interacting according to mixtures of various forces and subject to the randomization caused by temperature and quantum effects. There is no elixir of life, there is no life force, there is no mind that is not materially based, there is no soul. These attitudes are not in question among scientists, just as there is no question that we and potatoes evolved from a common ancestor. If either of these tenets—the molecular basis of our lives, or the idea that biological systems have evolved—were incorrect, then our whole agricultural industry, our medicine, our chemical industries, our pharmaceutical industries, our epidemiology, and our conservation efforts would all be based on incorrect assumptions and work as they do by pure fluke. There are still some details to be worked out about living systems, and no doubt there will be a few more intellectual leaps and dissonances over the next decade or two. They may be as disruptive to biology as was quantum mechanics to physics or computation to mathematics, but there will be no wholesale abandonment of our current understandings. The central tenet—that we are the product of trillions upon trillions of mindless molecular interactions and nothing more—will stand: It is not phlogiston or ether but a fact confirmed in thousands of new experiments every day of the week, every week of the year.

Most people have managed to remain blissfully unaware of the consequences of fifty years of molecular biology, and they are just now starting to take notice. Not too long ago, we saw the president of the United States on national television carefully parsing the subtle nuances of biological research as he announced his decision, based on ethical and political considerations, about what sort of stem cell research the government would fund. This was undoubtedly not the last time we would see our presidents so confounded. Or the last time we would see a confusing crossing of traditional lines by advocates on both sides of the argument. And there will certainly be more and more demonstrations, some of them violent—not just against genetically modified food, as there are now, but against technologies that will be seen as debasing us, as placing us in the same class as our manipulable artifacts.

We have already started to turn the analytical tools of molecular biology, developed over the past fifty years, into engineering tools. And with that, we are realizing our ability to manipulate life itself—and human life, in particular—at the most basic level of its operation.

Fifty years ago, just after the Second World War, there was a transformation of engineering. Before that, engineering had been a craft-based exercise, but starting around 1950 it was transformed into a physics-based discipline. Now we are seeing the beginnings of a transformation of engineering again, this time into a largely biologically based discipline, though it will not sacrifice the rigor of its physics background. At MIT's Artificial Intelligence Laboratory, where I am director, I see signs of this transformation every day. We have torn out clean rooms where we used to make silicon chips and installed wet labs in their place, where we compile programs into DNA sequences that we splice into genomes in order to breed bacterial robots. Our thirty-year goal is to have such exquisite control over the genetics of living systems that instead of growing a tree, cutting it down, and building a table out of it, we will ultimately be able to grow the table. We have turned labs where we used to assemble silicon and steel robots into labs where we assemble robots from silicon, steel, and living cells. We cultivate muscle cells and use them as the actuators in these simple devices, the precursors of prostheses that will be installed seamlessly into disabled human bodies. Some AI Lab faculty who study how to make machines learn have stopped building better Web search engines and begun inventing programs that can learn correlations in the human genome and thereby make predictions about the genetic causes of disease. We have turned rooms that used to house mechanical CAD (computer-assisted design) systems into rooms where we measure the cerebral motor control of human beings, so that eventually we can build neural prostheses for people with diseased brains. And our vision researchers, who used to build algorithms for detecting Russian tanks during the cold war, now build specialized vision systems to provide guidance during neurosurgery. Similar transformations are happening throughout engineering departments, not just at MIT but all over the world.

The first generations of these changes in perspective are rapidly accelerating the absorption of silicon and steel technology into our bodies. Early participants in these practices were driven by solid clinical reasons; they were trying to compensate for the injury or degradation of their bodies. We have long had pacemakers and artificial hips, and recently artificial hearts. But now more complex neural prostheses are becoming commonplace. Tens of thousands of people who suffered a profound loss of hearing have had devices permanently implanted in their cochleas; these implants provide a half dozen frequency bands as direct neural stimulation at a site in the cochlea that would be sensitive to that frequency in a healthy ear. These people hear through direct stimulation of their peripheral neurons by electronic circuits—more specifically, through a combination of silicon and "wet" neural circuits.

People who suffer from macular degeneration of the retina will be prime customers when a similarly effective visual implant is perfected. Teams around the world are working on ideas for implanting a silicon camera chip in the human retina and either connecting the picture elements directly to nerves in the retina or sending them, via cable or wireless, to the early vision-processing areas that lie in the back of the head. There have been a series of experiments using short-term implants of such devices, and at the time of this writing three patients have had retinal implants for more than a year—though there has not, as yet, been any publication of the results. Successful visual implants are much harder to achieve technically than cochlear implants, if only because there need to be thousands of accurately placed connections between the camera chip and neurons rather than the handful that work for conveying ordinary speech. However, there is every reason to believe that retinal implants will eventually become as routine as the cochlear variety.

A few quadriplegics whose injuries are so high up their spine, even into their brain stem, that they cannot speak or control their own breathing and need a respirator, are now able, using neural implants in their brains, to direct the mouse of a computer just by thinking. This interface allows them to communicate again with the external world and thus exercise some control over it. At the very least, they can choose what they want to see on their computer screens and somewhat laboriously type commands, messsages, and email. In some experiments, they can control robots that assist them in the tasks of daily living. One must assume that these sorts of techniques, which have begun to return some basic human dignity to the profoundly injured, will continue and that the scope and flexibility of the devices will improve over time.

There are many other experiments in introducing silicon and steel into the bodies of those with medical problems— from systems to exercise the muscles of stroke victims and victims of spinal injury to schemes to reroute neural signals in patients with Parkinson's and similar diseases. These and the experiments with quadriplegics have led to hope about the adaptability of our critical brain regions.

Before long, such clinical procedures will begin to be used in elective ways. Over the next ten to twenty years, there will be a cultural shift, in which we will adopt robotic technology, silicon, and steel into our bodies to improve what we can do and understand in the world. People who are not blind may choose to have a device sensitive to infrared or ultraviolet installed in one of their eyes. Or we may all be able to have a wireless Internet connection installed directly in our brains—although just what the Web pages we will browse with it might "look" or "feel" like in this mode is not yet known.

And then, approximately a quarter of the way through this century, similar enhancements of a more biological nature will begin to be available to us. In that time frame, the really large-scale use of genetic engineering will be common—beyond agribusiness and medicine, where it is currently being explored. Genetic engineering will be used in the petroleum industry, in the production of plastics and other materials, in recycling, in batteries, in renewable energy sources, and in other applications hard to imagine from this vantage point. By 2025 we will also have achieved enough explicit control to apply these technologies with confidence to our own bodies. This coincidence should be no surprise—it will have been enabled by the same science and technology applied in orthogonal directions.

Some of the early biological augmentations of ourselves may entail increasing the number of neurons in our cortex. Already these sorts of experiments are being carried out on rats. When extra layers of neurons are placed in the brain of a rat at a critical time in its development, its intelligence is enhanced relative to rats without this augmentation. As we better understand the hormonal balances that control the growth of our brain in childhood, we will perhaps be able to add sheets of neurons to our adult brains, adding a few points to our IQ and restoring our memory abilities to those we had when younger. There will likely be some errors and horror stories about augmentation gone haywire, but make no mistake—the technology, in fits and starts, will proceed.

By the midpoint of the twenty-first century, we will have many, many new biological capabilities. Some of them seem fanciful today, just as projections about the speed, memory, and price of today's computers would have seemed fanciful to the engineers working on the first digital computers in 1950. It seems reasonable to assume that by the year 2050 we will be able to intervene and select not just the sex of a baby at the point of conception but also many of its physical, mental, and personality characteristics as well, a much less trivial matter. We have seen how the ability simply to determine the sex of a fetus has badly skewed sex ratios in China and India; we can expect that these new capabilities will have profound and essentially unpredictable (at this stage) effects on the makeup of the world's population.

We will also be able to change already existing bodies. Surgical body modification and biochemical alterations (for example, through the use of botulinum toxin) have become common in the Western world in the last twenty years; fifty years from now, we can expect to see radical alterations of human bodies through genetic modification. Many of these modifications will surely be aimed at extending lifetimes, but many will be recreational and lifestyle-related. The human menagerie will expand in ways unimaginable to us today.

The techniques developed to enable these alterations of our bodies will also be used in our industrial infrastructure. Much of what we manufacture now will be grown in the future, through the use of genetically engineered organisms that carry out molecular manipulation under our digital control. Our bodies and the material in our factories will be the same. Perhaps we will be able to keep them separate in our minds, as we now mentally separate our circumstances from those of the chickens we raise in battery farms. But just as the shadow of those thoughts causes us to reflect on our own confined existences, there will be an alteration in our view of ourselves as a species; we will begin to see ourselves as simply a part of the infrastructure of industry.

While all the scientific and technical work proceeds, we will again and again be confronted with the same constellation of disturbing questions. What is it to be alive? What makes something "human"? What makes something "subhuman"? What is a superhuman? What changes can we accept in humanity? Is it ethical to manipulate human life? Is it ethical even to manipulate human life in particular "corrective" ways? Whose version of "corrective"? Whose version of "life" and "human"? What responsibility does the individual scientist have for whatever forms of life he or she may manipulate—or create?

And these questions will not only be asked in the well-meaning precincts of science; they will be thrashed out in the larger society, accompanied by everything from vandalism to terrorism to full-fledged war.

Our old generalizations—the initally uncomfortable ones of the last five hundred years—changed only our understanding of our position in the cosmos. Over the next fifty years, our new generalizations will empower us to change that position itself. We are breaking out of our roles as passive observers of life and the order of things to become manipulators of life and the order of things. We will no longer find ourselves confined by Darwinian evolution. Now we will have the option of participating in explicit ways in that evolution, both as individuals and as a species. Our adventures in nuclear fission will seem like child's play in comparison. We will need to observe and moderate our own hubris very carefully, if we want our descendants to someday be the objects of joyous discovery by a SETI effort somewhere else in our galaxy.

❑

RODNEY BROOKS is director of the Artificial Intelligence Laboratory and Fujitsu Professor of Computer Science at the Massachusetts Institute of Technology. He is also chairman and chief technical officer of iRobot Corporation, a company that partners with established companies in the toy, oil, consumer, and defense industries. Dr. Brooks appeared as one of the four principals in the 1997 Errol Morris movie *Fast, Cheap, and Out of Control*—named after one of his papers in the *Journal of the British Interplanetary Society*. He is the author of *Model-Based Computer Vision, Programming in Common LISP, Cambrian Intelligence,* and most recently *Flesh and Machines: How Robots Will Change Us*.

PETER ATKINS

❑

The Future of Matter

CHEMISTS ARE MAGICIANS with matter. They spin new
materials from the earth, the air, and the oceans and produce forms of matter that perhaps do not exist anywhere
else in the universe. However, unlike magicians, their practice is rational; they base their spinning on their deep
understanding of how atoms link together and can be
coaxed into new combinations. Their rational understanding of matter, the understanding that gives them such
power, emerged from experiments done in the eighteenth
and nineteenth centuries and was rendered quantitative by
the application of quantum mechanics to chemistry in the
twentieth. At the beginning of the twenty-first century,
chemists are in complete command of matter.

Chemistry spins in two directions. It spins new products
and it spins new subjects. The products spun by chemistry
during the past fifty years are all around us: the pharma-
ceuticals that render life longer and more agreeable and
death less painful; the fabrics and dyes that enliven the
everyday scene; the plastics and ceramics that displace
wood and iron, lightening and strengthening structures and
permitting the achievement of new and interesting shapes;

the semiconductors that have transformed society and the superconductors that we anticipate will transform society yet again; the fuels that enable us to live as we wish. But chemistry spins new subjects, too. These subjects take on separate names but are still essentially chemistry. Materials science is chemistry—chemistry applied to the generation of new materials with specific mechanical, electrical, and magnetic properties. Molecular biology, that extraordinary creation of the twentieth century and the foundation of biology and medicine for the twenty-first century, is chemistry applied to the molecules of formidable complexity that are responsible for life. Modern medicine, apart from the parts that hack, slice, and saw, is chemistry applied, in the past applied on a wing and a prayer but increasingly applied successfully with understanding. Anything that deals with the properties and transformations of matter is, at root, chemistry, regardless of whether that matter is living or dead.

The next fifty years will see the consolidation of chemists' ability to manipulate atoms and link them in novel patterns. There are three routes forward. One is the elaboration of the classical techniques of chemistry: the stirring, heating, and mixing that in various sophisticated forms have emerged from alchemy and have reached a high degree of refinement in our laboratories. In particular, organic chemists, who deal with compounds built primarily from carbon, have amassed a huge corpus of knowledge about how to link atoms in specific patterns, and such knowledge will undoubtedly be brought into the domain of intelligent systems. Synthetic strategies will be designed increasingly by computers that use neural nets to assess the best way forward. Computer-aided chemistry will become more and more essential, as chemists attempt the synthesis of ever more elaborate structures—not only proteins and nucleic acids but also organic materials for computation and data storage. In the next fifty years, we shall see computers designing synthetic pathways to intricate products needed for their own successors. Computers must get smaller. In due course they must be built from the smallest components possible; that is, they must be built from molecules, for nothing smaller can have the complex architecture that qualifies it as a structure. So chemists will build molecular computers, using the skills they have developed in building less intricate molecules.

Organic chemistry emerged from the study of compounds that were thought to be fabricated only by living organisms. The view that organic compounds could be produced only naturally was overthrown early in the history of chemistry, and the extraordinary vitality of organic chemistry is a witness to that overthrow. One consequence is the prospect of organic chemists synthesizing life from scratch. To do so, they need to be able to synthesize DNA, or an equivalent molecular data-storage system, and package it in artificial zygotic systems protected by synthetic membranes and equipped with synthetic systems of metabolism for powering the replication. At first, they will use a mixture of synthetic and natural components—a strip of artificial DNA introduced into a natural egg—but they can already synthesize most of the components needed to provide a fully artificial system, and within the next fifty years they will have made a number of viable synthetic proteins. I do not expect true organochemistry—the synthesis of living organisms and their properties—to emerge from current organic chemistry within fifty years, but this period will see the production of working proteins and a good synthetic approximation to cell membranes. By mid-century the bits and pieces of fully synthetic life will be in position, and the future will bring the pieces together. In the longer term there will be no need to stick with carbon, and the speculative dream of at least partial incorporation of silicon and germanium into living things and the generation of an entirely new kind of life will come true. Success in such activity will undoubtedly and properly give rise to profound questions of ethics; but any prospect of success is so far in the future that for the time being such problems need not be taken seriously.

Although organic chemistry has shed the view that organisms are essential for the production of its compounds, some molecules are so intricate that they can indeed be produced only by living organisms. Chemists will use organisms to produce such materials, and will effectively farm these organisms for their products; the term "agrochemistry" will take on a new significance. This milking of bacteria is already happening but will grow in importance as genetic engineering becomes more effective. There is no reason why bacteria or plants cannot be farmed for simpler molecules, too, such as hydrocarbons for vehicles and the raw products of the petrochemical industry. Those applications will become of great importance in the next fifty years, as our bank of stored hydrocarbons becomes depleted and we have to produce fossil fuel analogs.

The third approach to synthesis, after classical techniques and microbial farming, will build on the ability of chemists to manipulate single atoms. Already, atoms can be moved around on surfaces into preselected locations, and the future will see molecules being built atom by atom. Novel individual structures will be built that could not withstand the maelstrom of ordinary existence in air or in solution. It is hard to imagine that these techniques could be used on an industrial scale, but industry is becoming more sophisticated and such bespoke tailoring of molecules cannot be ruled out.

The concept of bespoke tailoring conjures up the prospects for nanofabrication and nanotechnology in general. Chemistry will undoubtedly contribute to the fabrication of minute components and find ways of making entities that currently have to be (in effect) carved from slabs. Molecular engineering is already producing analogs of mechanical components on the scale of single molecules, and the sophistication of these structures will increase as chemists find ways of nanofabricating cogs, wheels, axles, belts, bridges, and all the other paraphernalia of engineering. The early evangelizers of nanotechnology took the view that chemists could assemble microscopic versions of macroscopic machines: molecular wheels with molecular disk brakes running on molecular axles with molecular ball bearings. Most such speculations assumed that the properties of atoms could largely be disregarded or modified at will: that intermolecular forces could be weakened or strengthened, bonding tendencies could be ignored, and so on. It is much more likely that chemists will develop molecular analogs of macroscopic machines which take into account the actual properties of the constituent atoms rather than ignoring them. One engaging possibility is that bacteria can be genetically engineered to excrete whole cogwheels, pistons, and springs, or even whole machines, the excrement not necessarily being purely organic but including other elements, too.

There are currently high hopes (but few successes) for nanofabrication using carbon and boron nitride nanotubes. Already, strings of atoms have been inserted into carbon nanotubes to produce insulated wires one atom in diameter—wires that will enhance the extreme nanofabrication of computers. There is every prospect that the dimensions of specialized computers can be brought down to the size of a speck of dust and sprayed like an aerosol. After all, an ant's brain is hardly any bigger than that and yet achieves remarkably specialized activities.

Carbon nanotubes will also become of enormous significance in macroscopic structures, such as suspension bridges and domes. They promise enormous strength compared to their weight. It is not inconceivable that geodesic domes built from struts of pure carbon nanotubes and clad in sheets of pure diamond will one day provide habitats to protect us from our own ecological misdemeanors on this planet, shields for the resuscitation of deserts, and colonies on Mars or even in interplanetary space. Fifty years is perhaps just within the range of time over which carbon nanotubes will become an industrial product.

It must be admitted, however, that little new understanding of chemistry itself will emerge over the next fifty years. The subject is already highly mature, and there are unlikely to be many surprises in connection with its fundamental principles. That is not to say that chemistry is reliably predictive. One of the great surprises of the late twentieth century was the discovery of fullerenes—the soccer-ball-shaped carbon-60 molecule and its analogs, including carbon nanotubes—and although they were anticipated, no one took the anticipation seriously. Theoretical chemistry is very good at the rationalization of observations in terms of quantum theory and statistical mechanics; it is less good at prediction. Thus we can expect more surprises, but we can be confident that all such discoveries will fall within our current canon of understanding.

That is not to say that theoretical studies are useless or entirely academic. The use of computers in chemistry is already of great importance and will certainly become more important during the next fifty years. As their knowledge base grows, and as chemists increasingly use neural networks to guide them, and as the computation of bulk properties from individual molecular structures becomes more reliable, so computers will increasingly become infallible consultants. A major current application is the screening of compounds for pharmacological activity by computing the properties of their molecules and assessing whether those properties are promising. In principle, such screening can reduce the development time of drugs by years. There is already one worldwide screening computation in action which uses (like the current search for extraterrestrial intelligence) the downtime of networked computers throughout the world to scan the molecules that might be pharmacologically active. This use of computers will doubtless grow, especially since the successful virtual completion of the Human Genome Project has made available such a wealth of data.

Computers will increasingly be used in chemistry to guide the syntheses of other compounds, including catalysts—substances that enable particular reactions to proceed at appreciable rates without themselves being consumed. (The Chinese characters for "catalyst" also signify "marriage broker," which captures the sense very well.) They are the hormones of industry, and the chemical industry simply would not exist without them. The major research effort of most petrochemical industries is in the discovery and development of more efficient, cheaper, longer-lasting, and more selective catalysts. As long as industrial chemicals are produced, catalysts will need to be deployed. The early catalysts, such as lumps of iron or gauzes made of platinum and rhodium, were ostensibly quite simple, but catalysts are becoming increasingly sophisticated. Over the next fifty years, chemists will develop solid catalysts and new generations of homogeneous catalysts that dissolve in fluids and bring about their action in solution. Solid catalysts will increasingly take the form of microporous materials, which are solids pervaded by a labyrinth of molecular-size holes, tunnels, and cages. The great advantage of microporous materials is that they present a huge surface area (they are virtually entirely surface) and are highly selective to the types and sizes of molecules that can penetrate them. Computers are increasingly being used to discover the functions of these materials and design them afresh. The next fifty years will see a surge in the rational design and use of these materials and a flow of novel, cheaper materials from the industries that deploy them.

I have so far said nothing of the perhaps more traditional wing of chemistry's activity—the analysis of materials to discover what is present. The refinement of chemical analysis over the past fifty years has been due almost entirely to the development of three approaches: First, there is chromatography, in which substances move at different rates as they pass through a long thin tube. Then there is mass spectrometry, in which molecules are blasted apart and their identities inferred from the fragments they produce. Both techniques, which are of great sensitivity and often used in conjunction, will certainly undergo further refinement to enable them to identify ever tinier amounts of material. Third, there is the whole stable of techniques classified as spectroscopic, which involve monitoring the absorption of different kinds of electromagnetic radiation (ultraviolet, visible, infrared, microwave, and so on). Of these techniques, by far the most useful is nuclear magnetic resonance (NMR), which is also the basis of magnetic resonance imaging techniques used so effectively in medicine.

NMR has proved to be the most amazingly adaptable of chemistry's tools. In the early days of the technique, about fifty years ago, the procedure involved monitoring the absorption of radio waves when a hydrogen nucleus reversed its orientation in a strong magnetic field. Since then the technique has matured, as the electronics associated with it have become more sophisticated and whole groups of hydrogen atoms and other kinds of nuclei have been made to reverse their orientations collectively. The point I want to emphasize is that the technique has grown almost organically over the decades and shows every sign of growing in complexity for the next fifty years. It seems to reach a plateau of achievement, only to be reinvigorated by the addition of more complexity. With each increase in complexity, chemists can extract more information about the molecules in the sample; recent major successes are the determination of the structures of proteins in conditions resembling their natural habitat, the aqueous interiors of cells. Such is the adaptability of the technique that it is also being explored as a way of enabling the construction of quantum computers. Who knows—at some stage in the future, when quantum computing is achieved, an NMR spectrometer may even begin to think about the molecules it is studying!

Chemistry is about structure as well as composition. Chemists seek to understand the properties of molecules in terms of their shapes and sizes as well as of their arrangement of atoms. Thus they can trace many of the properties of water to the fact that its molecules are V-shaped, and they seek to understand the properties of proteins in terms of the helices, sheets, twists, and turns of these important molecules. Here, too, there are theoretical and experimental problems that we can expect to be solved in the next fifty years.

A theoretical problem currently attracting much attention is the following: Given the sequence of amino acids that go to make up a polypeptide chain (that is, the backbone of a protein molecule), what shape does the chain adopt in its natural environment? This is a crucial question in molecular biology, because the shape of a protein molecule effectively determines its function. Even disregarding the extraordinarily interesting pure knowledge that comes from tracing composition to function, the establishment of function can be seen as an essential component of the Human Genome Project, wherein we trace the information in DNA through to the proteins it encodes and then on to the functions they exercise by virtue of their composition and shape. One approach to the protein-folding problem (as it is called) is computational, but computers of enormous power are needed to analyze the contortions into which a long polypeptide chain can twist and then be trapped. This problem is slowly giving way and can be expected to occupy a great deal of chemistry and a large number of chemists over the coming decades. There is no point in synthesizing a polypeptide chain—a reasonably easy job—if it does not adopt the correct shape for the function it is to perform.

The experimental problem of the determination of shape has largely been solved by the introduction of X-ray diffraction. That technique is now a century old and reached its temporary apotheosis in the middle of the twentieth century with the determination of the structure of DNA and of a number of important proteins, such as lysozyme, insulin, and hemoglobin. Recent advances in the technique, which promise to be the basis of its development over the coming decades, are in the use of very intense X-ray sources such as are obtained from synchrotrons—huge devices in which electrons are constrained to move at high speeds in a circle, generating X rays as they change direction. Synchrotrons are national facilities and are coming into use in a number of centers around the world; the high-intensity X rays they produce enable X-ray diffraction patterns to be obtained much more rapidly and with much more detail. We shall start to see the determination of the structures of molecules in solution and perhaps even the observation of reactions in progress.

Synthesis, analysis, and structure are three of the major components of chemistry, and we have touched on them all. Finally, there is reaction, the actual process by which one substance is changed into another. Recent advances in spectroscopy, relying on the use of pulsed lasers, have enabled chemists to examine reactive events on timescales measured in femtoseconds (10^{-15}s, or one thousand-trillionth of a second). On this timescale, an atom in flight is hardly moving. So far only very simple reactions have been examined on such a short timescale, but it is possible to envisage developments of the technique in which real reactions, perhaps even the reactions catalyzed by enzymes, can be examined in this way. Then we shall have frame-by-frame movies of reactions in progress and watch atoms and molecules in the most intimate moments of their existence, giving us at last true and deep insight into the forms of matter that we manipulate with our magical techniques.

❏

PETER ATKINS is a professor of chemistry at the University of Oxford and a fellow of Lincoln College. His research has been in the field of theoretical chemistry, particularly magnetic resonance and the electromagnetic properties of molecules. Nowadays he spends virtually all his time writing; he is the author of several textbooks *(General Chemistry, Physical Chemistry, Inorganic Chemistry, Molecular Quantum Mechanics, Quanta, Concepts of Physical Chemistry)* and books for general audiences such as *Molecules; The Second Law; Atoms, Electrons, and Change;* and most recently *The Periodic Kingdom.*

ROGER C. SCHANK

❑

Are We Going to Get Smarter?

Is INTELLIGENCE AN ABSOLUTE? Does mankind get smarter
as time goes by? It depends on what you mean by intelligence, of course. Certainly we are getting more knowledgeable. Or at least it seems that way. While the average child
has access to a wealth of information, considerably more
than was available to children fifty years ago, there are
people who claim that our children are not as well educated as they were fifty years ago and that our schools have
failed us.

Today, questions about what it means to be intelligent
and what it means to be educated are not at the center of
our scientific inquiry, nor are they at the center of our popular discourse. Still we live our lives according to implicitly
understood ideas about intelligence and about education.
Those ideas will be seriously challenged in the next fifty
years.

About ten years ago, I was asked to join the board of
editors of *Encyclopaedia Britannica*. The other members
were mostly octagenarians and mostly humanists. Because
I was both a scientist and much younger than everyone
else, most of what I said was met with odd stares. When I

asked the board if they would be happy to put out an encyclopedia ten times the size of the current one if the costs involved remained the same, they replied that, no, the current encyclopedia had just the right amount of information. I responded that they would be out of business in ten years if that was their belief. They had no idea what I meant—although I tried to explain the coming of what is now called the World Wide Web. At a later meeting, after having heard me make similar assertions about the future, Clifton Fadiman, a literary hero of the 1940s, responded, "I guess we will all have to accept the fact that minds less well educated than our own will soon be in charge of institutions like the Encyclopaedia."

The chairman of the board of the *Encyclopaedia Britannica* at that time was the late Mortimer Adler. He was also responsible for a series called The Great Books of the Western World, which was (and is) sold as a set. These books represent all the great written works of the world's wisdom—according to Adler and his colleagues, anyhow—and the series consisted mostly of books written prior to the twentieth century. I asked Adler whether he thought there might be some new books that could be included, and he replied that most of the important thoughts had already been written down.

This idea, that all the great thoughts have already been thought, has been prevalent in the American idea of education and intelligence for a long time. Here are the admission requirements for Harvard College in 1745:

When any Schollar is able to read Tully or such like classicall Latine Authour *ex temporare*, and make and speake true Latin verse and prose *Suo (ut aiunt) Marte*, and decline perfectly the paradigms of Nounes and verbes

ine the Greeke tongue, then may hee bee admitted into
the Colledge, nor shall any claim admission before such
qualification.

What the Great Books series and Harvard of 1745 have
in common is an underlying assumption that the study of
man and his institutions had been sufficiently mastered in
ancient times and therefore education required you to be
well read and well versed in the thoughts of those who had
preceded you. An educated person in this view is one who
is able to discuss with erudition a variety of historical,
philosophical, and literary topics. Being educated—and
therefore being intelligent—has, for the last century and
many centuries before that, been about the accumulation
of facts, the ability to quote the ideas of others, and a familiarity with certain ideas. Education has meant accumulating information, and intelligence has often meant little
more in the popular imagination than the ability to show
off what one has accumulated.

But what happens when the facts are in the walls?

Fifty years from now, knowledge will be so easy to
acquire that one will be able simply to say aloud whatever
one wants to know and hear an instantaneous response
from the walls—enhanced by a great deal of technology
inside those walls, of course. Knowing offhand what Freud
had to say about the superego won't mean much when you
can turn to the nearest appliance and ask what Freud had
to say and and hear Freud (or someone who looks and
sounds a lot like him) saying it and finding five opposing
thought leaders from throughout time ready to propose
alternative ideas if you want to hear them and discuss
them together.

But is intelligence simply the ability to be informed of answers to your questions, or is it the ability to know what questions to ask? As answers become devalued, questions become more valued. We have lived for a very long time in an answer-based society. Signs of it are everywhere: in the television shows that people watch, such as *Jeopardy* and *Who Wants to Be a Millionaire?*; in the games that people play, such as Trivial Pursuit; and most of all in school, where answers are king. Increasingly, the chief concern of our schools is testing. School has become a regimen for learning answers rather than learning to inquire.

New technologies will change all this. When the pocket calculator was introduced, people asked whether calculators might as well be used in math tests, since from now on such devices would always be available. As a result, math tests began to focus on more substantive issues than long division. The introduction of artificial intelligence into everyday devices will have the same effect. As machines become omnipresent and able to answer questions about whatever concerns us, the values we place on each individual's being a repository of factual knowledge will diminish. The old idea of school, based on the notion that the most knowledgeable person in town had information to impart and the rest of us were forced to sit and memorize that information, will give way to new ideas of knowledge acquisition. Knowledge will no longer be seen as a commodity to be acquired. Anything obtained easily is devalued in society, and it will be the same with knowledge.

What will be valued will be good questions. *Computers can only take you so far*, we will hear people say.

Imagine the following: You are sitting in your living room, talking with your spouse, and an issue comes up between you. You turn to the wall for a response. "Who was right?" you demand. The wall points out that it has a number of virtual people available to join your conversation. You choose some characters whom you have heard about or conversed with before. A lively discussion ensues. Eventually the limits of the computer's collective knowledge are reached. The walls know no more of relevance. "This, then, is an exciting question!" you exclaim. Knowing a good question makes you ready to enter into a discussion with other live humans interested in similar questions. You tell that to the walls, and suddenly the people interested in such questions— those who have gone beyond the software in the same way that you have—are all there in your living room (virtually). In a world where this is possible, what does it mean to be educated? What does it mean to be intelligent?

To think about the education part of that question, we have to ask what a child's life would be like in that world. Fifty years from now, school as we know it will have atrophied from lack of interest. Why go to school to learn facts, when virtual experiences are readily available and the world's best teachers are virtually available at any moment? Education will mean—even from the age of two—exploring worlds of interest with intelligent guides available to answer your questions and pose new ones. World upon world will open to the child who is curious. Education in such a society will be a matter of what virtual (and later real) worlds you have entered and how much you have learned to do in those worlds.

To Fadiman's remark quoted above, I responded that minds would not be less well educated, just differently educated. In the world of Clifton Fadiman, an educated mind was one that had been trained at Harvard (or its equivalent) and was conversant with the major ideas in Western thought. His idea of education did not include, for example, being able to program in JAVA, or understanding the basics of neuroscience. In fifty years, there will still be Harvard, but the value of its imprimatur will have been altered tremendously.

Education in its deepest sense has always been about doing, rather than about knowing. Many scholars throughout the years have pointed this out, from Aristotle ("For the things we have to learn before we can do them, we learn by doing them") to Galileo ("You cannot teach a man anything; you can only help him discover it within himself") to A. S. Neill ("I hear and I forget; I see and I remember; I do and I understand") to Einstein ("The only source of knowledge is experience"). Nevertheless, schools have ignored this wisdom and chosen—in the words of John Dewey—to "teach by pouring in."

The virtual schools that will arise to take the place of current institutions will attract students less because of the credentials they bestow than because of the experiences they offer. Since these experiences will be there for the taking when a learner decides to learn, most students will start college long before the age of eighteen. Success in various virtual experiences will encourage us to encounter new ones, much as video games do today. Certifying agencies will worry more about what you can do—what virtual merit badges you have achieved—than what courses you have taken.

Fields of endeavor will create experiences in those fields. Instead of Harvard or Columbia offering courses in physics, physicists from around the world will work with virtual-educational-world designers who will build software to create physics experiences. Those experiences will be available to everyone. The old idea that the smartest people were those who received the best grades from schools that tested them to see how well they had learned the lessons will morph into a notion that the smartest students are the ones who pose questions for the software that have to be sent to humans in order to be answered. Intelligence will mean the ability to reach the limits of an educational experience.

Will we collectively be smarter as a society because of all these innovations? In terms of raw capacity for thought, people are as smart now as they ever were or ever will be. But a brilliant cave dweller, who had available to him limited knowledge of the world and limited wisdom from the ages, could work only within the parameters of the tools he knew. He may have understood the nature of humans and their institutions as well as the Greeks who followed him. He may have been as intelligent as the Greeks who followed him. But in any absolute sense he wasn't too smart, because there was so much that he hadn't experienced.

The same is true of our view of the Greeks, of course. Aristotle seems brilliant because he tackled issues we still tackle today and had great insights into those issues. Yet Aristotle can also be almost funny in his naïvete when he approaches subjects with which he had little experience and with which we have had so much more. Each generation improves on the experiences it opens up to the next. But a leap of tremendous proportions is coming in the next generation. The fact that we still have teachers and classrooms and textbooks will be almost laughable in fifty years. People will look back at us and ask why it took so long for us to change our notions of education, why we thought SAT scores mattered, or why we thought memorizing answers was a mark of intelligence in any way. The notion that education is about indoctrination by the state—an idea boldly stated in the 1700s and little acknowledged today—will seem scary. The governmental control of information—still popular in some countries, and still possible in those countries without computer access—will become an archaic notion. Too much experience will be available too readily and too cheaply to prevent anyone from experiencing anything. Governments will have to give up even imagining that they are in the education business, an area they dominate today, and will be unable to control the broad distribution of virtual experiences in much the way that they are failing to control television and computer access in country after country today.

We will begin to understand in the next fifty years that experience and one's ability to extend its range is the ultimate measure of intelligence and the ultimate expression of freedom. The creation of virtual experience will become a major industry; our homes will be dominated by virtual experiences; our schools will have been replaced by them. What we see today in video games and science fiction movies will become our reality. Today, games like Everquest attract hundreds of thousands of players, who inhabit virtual worlds in an effort to gain status, form relationships, and acquire various virtual objects. These games are so real to the participants that the virtual objects they employ are for sale (for hefty prices) on e-Bay. Many players of these games have a social life entirely based upon them. In the future, these worlds will become much more sophisticated and even more intertwined with the real world.

We really will be able to go wherever we want to go on any given day, and all anyone will ask of us is where we have been and what experiences we have had there. We will seek out those who are more experienced than us in the virtual worlds they have entered. We will understand that it is the questions that remain unanswered and those who can think critically about them that are the factors in any true measure of intelligence. Of course, this last idea is well understood in universities today, but it is not really appreciated in business or government. Politicians want simplistic points of view, teachers want correct answers, businesses want solutions, venture capitalists want profits, the media want national soap operas, certifying agencies want scores. Those who are considered smart in a society like that are those who have succeeded in supplying it with what it wants. In such a supply-and-demand view of knowledge and intelligence, even Clifton Fadiman would have felt left out. Still, he and those of his generation could hold themselves above all this and talk about Great Books.

I was once asked to review some technical colleges to see how they were teaching. In a class for future chefs, each student had his own cooking facilities and they were busy making food. All I could say was that I had nothing of interest to add. The school was teaching doing by having students do. While this is not a radical idea in technical colleges, it seems to be radical in our other institutions of higher learning. As more tools for doing become available, it is doing that will matter. At Carnegie Mellon, where I work, new students must put together their own computer as soon as they arrive on campus and use that computer for the next four years. You can be sure that they understand how computers work once they have built one themselves.

It is what we can do, not what we know, that will matter in an educational system based on realistic performance environments. The important intellectual issues will revolve around questions arising from the nature of students' interactions in the virtual educational world.

When educational environments demand questions, ask how questions were obtained, and demand to know the experiences that brought on those questions, then the profound change that computers offer will have been realized. We will all be smarter—a great deal smarter—in the sense that we will not be afraid of new experiences. We will know how to find those experiences and we will grow from them. Our minds will be differently educated and our intellectual world will be dominated neither by humanists nor by scientists but by experientialists, those who have been there and have become curious as a result.

❑

ROGER C. SCHANK, a leading researcher in artificial intelligence, is the chairman and chief technology officer for Cognitive Arts and Distinguished Career Professor in the School of Computer Science at Carnegie Mellon. He was formerly the director of the Institute for the Learning Sciences at Northwestern University, where he is professor emeritus. His books include *Dynamic Memory: A Theory of Learning in Computers and People*; *Tell Me a Story: A New Look at Real and Artificial Memory*; *The Connoisseur's Guide to the Mind*; *Virtual Learning: A Revolutionary Approach to Building a Highly Skilled Workforce*; *Coloring Outside the Lines: Raising a Smarter Kid by Breaking All the Rules*; and *Designing World Class E-Learning*.

J A R O N   L A N I E R

❏

The Complexity Ceiling

THE FIRST FIFTY YEARS of general computation, which roughly spanned the second half of the twentieth century, were characterized by extravagant swings between giddy overstatement and embarrassing near-paralysis. The practice of overstatement was initiated by the founders of computer science: Alan Turing wondered whether machines, particularly his abstract "universal machines," might eventually become the moral equivalents of people; in a similar vein, Claude Shannon defined the term "information" as having ultimate breadth, spanning all thermodynamic processes.

One could just as well claim that since all life is made of chemical interactions, any chemical apparatus can be understood as a nascent version of a person. The reason this claim isn't made is that the difference in complexity between the chemistry of living things and what can be studied in contemporary chemistry laboratories is apparent. We have intuition about the distinction. In contrast, we do not have a clear intuition about the differences in complexity between the various kinds of information systems. A serious and intelligent community of researchers who describe themselves as studying "artificial intelligence" believed, in some cases as early as the late 1950s, that computers would soon become fluent natural-language speakers. This hasn't happened yet, of course, and we still don't have an intuition of how large a problem it is to understand natural languages, or how long it might take to solve.

The practice of overstatement continues, and it is even common to find members of elite computer science departments who believe in an inevitable "singularity," which is expected sometime in the next half century. This singularity would occur when computers become so wise and powerful that they not only displace humans as the dominant form of life but also attain mastery over matter and energy so as to live in what might be described as a mythic or godlike way, completely beyond human conception. While it feels odd even to type the previous sentence, it is an accurate description of the beliefs of many of my colleagues.

Some readers will note that I have been accused of similar overstatement in regard to the term "virtual reality." What is often misunderstood, however, is that the goal of virtual reality is not to thoroughly describe and reproduce physical reality (a project that might well be impossible) but to understand human cognition well enough to engage the human nervous system in an evolving game of illusion. Virtual reality is essentially the scientific study of the limits of stage magic, not of the reduction of physical reality.

Accompanying the parade of quixotic overstatements of theoretical computer power has been a humiliating and unending sequence of disappointments in the performance of real information systems. Computers are the only industrial products that are expected to fail frequently and unpredictably during normal operation. The underestimation of the expense of maintaining information systems is almost a constant; it might even be called a ritual of contemporary business.

Specifically, it is the software that seems impossible to manage for a predictable price. And only certain kinds of software. Hardware continues to get smaller, faster, and cheaper at the exponential rate known as Moore's Law, and it is this breathless success that fuels the fanatic overstatement. Software in closed systems, with interfaces that are small and constant enough to be specifiable, can also be made reliably, though not cheaply. An example of this kind of software is the code that runs a modern aircraft, such as an Airbus. The kind of software that doesn't seem to be manageable has a complicated and changing interface to its surroundings. An example is personal computer software, which is notoriously unruly. It is important not to confuse the two kinds of software. The end of the twentieth century saw an odd and widespread paranoia that Y2K bugs would bring about a massive disruption. The reason this did not happen is that most infrastructure software is of the type that is manageable, albeit at gargantuan expense.

One possibility for the next fifty years of computer science is that the same two trends—overstatement of potential and underestimation of expense—will simply continue. This is a scenario that might be termed "Planet of the Help Desks," in which the human race will be largely engaged in maintaining very large software systems. It is not an entirely unappealing prospect, since it would keep humanity gainfully employed. This dull future is not inevitable, however, and it is worth imagining a new phase of computer science that will offer fundamentally fresh possibilities.

As a start, computer science must go back to its origins and rethink the relationship of information to physical processes. Claude Shannon made the brilliant conceptual leap of connecting measurable bits to the entropy of a physical system, but this formulation in isolation is misleading. Not all bits can be measured in practice, and therefore some bits are more important than others. Most potentially measurable bits in a physical system are in practice lost in a sea of statistical distributions. A popular trope in the late twentieth century held that the beating of a butterfly's wings could be the ultimate cause of a thunderstorm weeks later on the other side of the earth. One problem with this idea is that even if it's true once in a while, there aren't enough storms to account for all the many butterflies. Measurable bits might be said to have different "causal potentials." Perhaps Shannon's information should be renamed "potential information." In order for a bit to be important— that is, to have high causal potential—it must be read; it must be a critical part of a system. This brings up what is sometimes called "semantics," or the context in which computation can be meaningful.

There has always been an (only occasionally acknowledged) observer problem in computer science. One way to state this problem is to consider an alien race that has no information about human language, history, or culture. Such aliens would no more be able to reliably reconstruct the meaning and function of, say, a personal computer in isolation than they would a lone edition of Shakespeare floating in interstellar space.

This is not a remote theoretical issue but an immediate practical one. Since the complexity of software is currently limited by the ability of human engineers to explicitly analyze and manage it, we can be said to have already reached the complexity ceiling of software as we know it. If we don't find a different way of thinking about and creating software, we will not be writing programs bigger than about 10 million lines of code no matter how fast, plentiful, or exotic our processors become.

At the dawn of computer science, in the mid–twentieth century, the only available intuition-building experience of information was the sending of pulses down wires. The early versions of information theory, which still dominate the standard curriculum, were concerned with single-point sampling of the world at the end of a wire. Therefore computer architecture as we know it was designed around simulated wires. Source code is a simulation of pulses that can be sent sequentially down a wire—as are passed variables, or messages.

The way to make pulses on a single wire meaningful is to have a protocol that assigns meaning according to sequence. Most of the first half century of computer science was inspired by such protocols. There have certainly been successes, such as the protocols that enable the Internet. But this is not the way natural systems work. While it might be theoretically possible to use twentieth-century-style algorithmic protocols to explain what the vision cortex does with signals received from the optic nerve, to do so would involve complexity on an entirely untenable scale. Clearly protocol adherence is not an efficient means of explaining a system that receives a large number of inputs in parallel, and it is also probably an inadequate method of engineering very large systems. If we replace the concept of a wire with that of a surface that can be sampled at many points, we must move away from algorithmic protocols and toward a new set of techniques, including pattern classification and the automatic maintenance of implicit confirmatory and predictive models.

Here is a current practical problem that illustrates this idea: For many years I have worked with surgeons to build simulations that will help them plan procedures for specific patients. The simulations are complex by contemporary standards and each one is built and maintained by teams of highly skilled specialists over periods of many years and must be tested with thousands of patients in order to become usable.

Now suppose that a team at one medical school has worked for ten years on building a brilliant virtual heart that has been shown to be usefully predictive of surgical outcomes. Meanwhile at another school there is a similar team that has invested a decade in a virtual lung. And suppose these two teams wish to combine their works into a virtual chest.

Almost certainly the two groups will be using incompatible protocols. Not only are they likely to have chosen different basic machines, operating systems, implementation languages, and the like, but there are probably also differences in conceptual approach. One group may have emphasized global, top-down, constraint maintenance, while the other may have been more partial to bottom-up rules. One group may have emphasized object semantics, while the other may have attempted an approximation of a continuous system. The current state of the art would have these two groups agree on a protocol of signals that could be sent down a wire between them. Such protocols are problematic; in this case the complexity might simply be prohibitive. We will know within a few years, since it is being attempted. If an interorgan protocol turns out to be possible, building one will force a tragic trade-off in the art of organ simulation. A working protocol would almost certainly reduce the prospects for improving any of the constituent organ simulations it connects.

To understand why this is so, we need to delve into the problem of legacies in information systems. The adjective that best describes our current software is "brittle." It breaks before it bends. This is the result of the emphasis on protocol adherence, which is an unforgiving requirement. Because of this underlying brittleness, software builds up in layers; it would be impossibly complicated and expensive to exhume protocols that have already been relied upon by many users in different ways. Thus we have the phenomenon of "lock-in," in which some software becomes effectively mandatory. Lock-in was manipulated in the late twentieth century by software vendors to build some of the greatest fortunes of all time.

Beyond lock-in is an even more annoying software characteristic, which I've dubbed "sedimentation." Software sedimentation is a process whereby not only protocols but the ideas imbedded in them become mandatory. An example is the idea of the file. Before about 1984, there was a controversy about whether files were a good idea. Some computer scientists felt that it would be better to have a finer-grained structure for shared information—in effect, a single global file made of tiny elemental components like letters. Indeed, the first internal version of the Macintosh computer didn't use files. Not only did the released version of the Macintosh have files, but so did Windows, Unix, and several other widely adopted systems. Files are now taught to students as a fact of life as fundamental as a photon, even though they are a human invention.

Back to the virtual heart and lung. As soon as the engineering groups have signed on to a protocol, the protocol becomes their master, since the groups would have to change simultaneously in order to revise it, and that would be effectively impossible because of the expense and complexity of the task. Whatever ideas about interorgan communication are in vogue at the time of the protocol's invention will be sedimented into place. Thinking will stop.

Therefore, a fine goal for the next fifty years of computer science would be to find an alternative to protocol adherence as a means of connecting components of large systems. In the case of the heart and lung, an alternative can already be glimpsed.

Suppose that each organ pretended that the other was a real, physical organ being probed by real sensors. Each organ could measure fundamental properties in the other, such as temperature, pressure, and chemical constituents at points in time and space. Each organ would appear to the other as a surface that could be sampled to varying degrees, but no higher-level parameters would pass between them. There would be no protocol other than the low-level one dictated by the nature of possible physical measurements.

For this scheme to work, each team would have to learn to recognize patterns in the other's simulation. The heart would no longer be able to send a message that it had performed a beat. That would have to be inferred by the lung from such things as fluid motion and tissue displacements. Each team would also learn to build a model of the other organ, in order to assist in interpreting measurements. These models might not exist as independent, separable structures but might be implicit in the chosen signal-processing methods, and would almost certainly be able to retune themselves with use.

This formulation might be called "statistical surface binding." If it can work for organ simulations, it might also work for general computer architectures. Perhaps in the future there will be an operating system whose components recognize, interpret, and even predict each other. Such a system would be less prone to catastrophic failure. There is no way at this time to know how well this sort of scheme might work, but some kind of statistical binding may well have to be adopted if computer architectures are to grow beyond the size we currently know how to manage.

As things stand, we have a way of thinking about an infernally tedious, low level of description of an information system (protocol adherence), and we have some lofty bird's-eye views of complexity from a purely theoretical perspective. But we lack an in-between perspective—a way to formally understand complex systems in terms of the relationships between large components. If we can model the human body from a surgeon's point of view as a graph of information surfaces, might the technique be generalized to other problems in understanding living systems?

Because we've had no intuition about the relative scales of information structures, we've had a hard time comparing our computational accomplishments to what nature has accomplished. Both the technical and popular press are awash with claims that human computational prowess is about to catch up with natural complexity. Examples include the repeated claims that computers are about to finally understand human emotions or language, or that computers are about to allow us to bridge the gap between complex organisms and the simple sequences of DNA we have learned merely to catalog.

One way to frame the nature of our ignorance in this matter is to ask whether natural evolution was a bumbling, slow, inefficient process or the result of a naturally self-

assembling parallel supercomputer (perhaps even operative on a quantum level in some cases) that self-optimized
to bring about an irreducibly complex result in roughly the
shortest possible time. These two alternatives are the outer
bounds of what could be true. The truth, which we don't
know, is somewhere in between. My bias is toward the latter bound: Evolution was probably pretty efficient at performing an irreducibly difficult task. It seems, however,
that the other extreme—that all it will take is another thirty
to fifty years of Moore's Law magic and our computers will
outrun nature—is accepted in most contemporary dialog
about the future of science and technology.

Wire- and protocol-limited mid-twentieth-century computer science has dominated the cultural metaphors of
both computation and living systems. For instance, Jorge
Luis Borges described an imaginary library that would
include all the books that ever were or possibly might be
written. If you happened to be lucky enough to live in a
universe big enough to contain it (and we aren't), you'd
need to invest the lives of endless generations of people,
who would wither away on starships trying to get to the
right shelf. It would be far less work to learn to write
good new books in the traditional way. Similarly, Richard
Dawkins has proposed an infinite library of possible animals. He imagines the invisible and blind hand of evolution
gradually browsing through this library, finding the optimal
creature for each ecological niche. In both cases, the authors
have been infected by the inadequate computer science
metaphors of the twentieth century. While an alternative
computer science is not yet formulated, it is at least possible to speculate about its likely qualities.

A new computer and information science would incorporate a theory of legacy. The configuration spaces of complex

causal systems are so vast that they cannot be understood as infinite libraries, because there would never be enough time or energy for useful browsing. Stuart Kauffman is fond of pointing out that our universe isn't old enough to have allowed for the exploration of all the possible proteins of even a reasonably small size, for instance. So complex systems accumulate legacies, which restrict the size of further configuration space searches. We must learn to give up the illusion that we can overcome legacies. This is the illusion in play when otherwise well-informed technologists propose radical additions to the human metabolism or brain structure (and yes, there are many such proposals).

One idea worthy of investigation is whether "legacy" is the same thing as "semantics." "Semantics" is a word that has been used to describe whatever mysterious thing lies beyond the syntax barrier characteristic of protocol-based systems: For instance, natural-language systems are always said to be progressing but lacking in their understanding of semantics. A legacy creates an immutable context in an information system. Legacies are complex. Legacies, in reducing the configuration space of a system, act like lenses that enhance the causal potential of bits.

Saying "I do" at a marriage ceremony is of more consequence than saying the same thing to a stranger who stops you on the street to ask if you have a match—at least it usually is. The marriage ceremony is a legacy, a pattern with a history that is expensive to undo. Similarly, DNA takes on meaning only in the context of an embryo; an isolated strand of DNA would almost certainly not be informative enough for our clever aliens of the earlier thought experiment to re-create a creature.

A new computer science might usefully incorporate a coarse-grained way of understanding natural systems as information systems—in a way that moves beyond the tiny-grained example given by Claude Shannon. It has often been stated that at the end of the twentieth century our understanding of physics is adequate to explain all the isolated events, such as chemical binding, that occur in living systems, so that we now must proceed to an understanding of complex systems. This is easy to say but hard to do. We must learn to parse natural systems according to causal potential. At any one time, only a small part of the material or energy of a system will significantly affect the future of that system—particularly if it's a living system. And even then there are differences in degree: A tiny change in a synapse can mean much more than a similar change on the surface of a skin cell or a cell almost anywhere else in the human body, for instance.

Stuart Kauffman has proposed that life might be defined as a process that both self-reproduces and performs a Carnot work cycle (the classic model for turning energy into work). This suggests at least one possible way to parse natural systems. Each Carnot work cycle is associated with a governor of some kind, a portion of the system that is responsible for restarting the cycle. These governors would have more causal potential than other material in the cycle; that is, the system could be disrupted by a smaller change to the governor than would be the case with other material in the same system. Whether this method of parsing is useful for coarse-grained understanding of the natural world as an information system remains to be seen, but some method should be found.

If we develop a formal and general coarse-grained method for parsing physical systems into causal information structures, we may well come up with a measure of complexity that incorporates both a counting and an energetic component. For instance, we could ask how expensive it would be for a system to probe its own internal causal structure, since the links in the causal chain could be physically characterized. Experience with coarse-grained explanations of simple natural systems might then prepare us to model the legacies that nature has evolved. In fifty years, if we're lucky, we might be able not just to describe how DNA works and what DNA is present (as we are beginning to now) but to have a way of describing the intermediate levels of complexity within which changes to DNA are constrained. In effect, we might learn to see the world to some degree from evolution's point of view, instead of from a molecule's or an organism's point of view.

In fifty years, biology and medicine will become a little like geography as we know it today. These fields of study will be mostly mapped out and much less mysterious. Alas, knowing how to map the earth can speed our travel between two points only to a limited degree. In the same way, being able to explain what are now mysterious aspects of biology will not automatically put them under our control. Instead, we're likely to discover which aspects of biology are irreducibly complex. There might be a reason that evolution took such a long time to attain certain configurations, and we might find that no shortcuts are available to us. This possibility is only one example of the far and final shore of the information science frontier. In a wide variety of explorations, from economics to agriculture, we will be limited by complexity ceilings—barriers that will not necessarily be breached by building bigger and faster computers. We will start to perceive complexity ceilings as the truest constraints of our possibilities. We don't yet know where they lie, but in fifty years we will.

❑

JARON LANIER, a computer scientist and musician, is best known for his work in virtual reality. He is the lead scientist for the National Tele-Immersion Initiative, a consortium of universities studying the implications and applications of next-generation Internet technologies.

DAVID GELERNTER

❏

Tapping into the Beam

WHAT WILL HAPPEN TO computer technology over the next fifty years? Where will we be half a century from now? Today, information is moving online; before long, the "great rationalization" will begin. Like a whole series of other technology industries (railroads, cars, radio, TV), the information industry will develop new standard forms. These forms will have nothing to do with today's commercial software applications. What matters is how the information itself is arranged; it's not the Web browser standard but the Web site standard that will count. (The Web itself will be obsolete, but the basic idea holds.)

The standard shape of information will be a form I'll call the *information beam*. The beam will be as important as the book. It won't replace the book; it will provide a comparably robust, sturdy, simple structure for the cyberworld. It will reshape our cultural life. Today we spend around 80 percent of our energy worrying about form (in many different ways) and 20 percent on content. Fifty years from now, this proportion will be reversed.

The most important information is fresh-off-the-wire, real-time information—information that tells us what is happening somewhere or other right now. Today "some-where or other" means someplace on Earth—the office, the school, the Senate, downtown. Before long, it will mean someplace in the Cybersphere. The continuous, ubiquitous Cybersphere will replace today's chaotic, stuttering Inter-net. The New York Stock Exchange, for example, is currently making the transition from place-on-Earth to place-in-the-Cybersphere; within the next half century, virtually all other institutions will follow. (This is a claim I first made in *Mir-ror Worlds*, in 1991, and I stand by it.) Without stirring from your favorite comfy chair, or getting out of bed, you will be able to tap into ongoing life at your office if you work, at school if you study, in the market if you shop, in the world if you care. But you won't stop venturing out. Instead, the physical and social landscapes will rearrange themselves to accommodate this new cultural state of affairs.

Before I sketch out these big changes, here are the nat-ural laws that will guide them:

1) *In the technology world, software, not hardware, deter-mines the state of the art and the pace of change.* The rate at which technology moves forward doesn't depend on the circuits (or computational proteins et al.) that we invent. It depends on the software architectures we design. If you dream up a new way of arranging infor-mation—a new software architecture—hardware will eventually be developed to support it. If you build pow-erful new hardware, in itself it will be absolutely useless. Ten seconds ago, my secretary walked in with another new future-of-computing book scouting for blurbs. Like the vast majority of such books, it is about hardware. Future-of-software books are vanishingly rare. No one seems to know the future of software.

When we contemplate what our technology will be like half a century from now, we're apt to think about blazingly fast circuits, molecular and optical computers, new data-transmission media, and other hardware miracles. These are important, amazing, and in themselves irrelevant. The shape of technology in half a century will depend on the software we invent.

Case in point: Since the mid-1980s, computing hardware has zoomed forward at enormous speed. And so what? Exactly how are you better off computationally than you were in 1985? Your bread-and-butter dealings with your computer are just what they were sixteen years ago. Your 2001 word processor is no better than the 1985 model; it eats up hundreds of times more memory and computing power, but it doesn't *do* anything differently (nothing important, anyway). Your spreadsheets are basically the same. Your email is the same—a lot more people use email, but email itself is the same as it was in 1985. Your desktop, file system, graphical user interface (if you owned a Mac in '85)— they are all just what they were a decade and a half ago. The only big change in your computational quality of life is the Web, but the Web is made of software, not hardware.

Today, software is stalled; therefore the technology industry is stalled. Hardware revolutions are a dime a dozen. To unstall things, we will need a revolution in software—and, bet on it, we will get one.

2) *The impossible-to-learn Law of Replacement: Society replaces a thing when it finds something better, not when it finds something newer.* Don't expect everything to be dif-

ferent fifty years from now. The fundamentals stay the same. This seems obvious, but isn't.

A front-page headline last summer in the *New York Times*: "FORECASTS OF AN E-BOOK ERA WERE, IT SEEMS, PREMATURE." A year earlier, in August of 2000, Barnes & Noble, Microsoft, and several other companies formally announced the E-book's arrival. Their big predictions were wrong. Is it conceivable that the book has lasted two thousand years because books are good and not because computer engineers haven't quite got around to replacing them? As I wrote in 1999 (also in the *Times*), "Replacing books by computers is like replacing cut flowers with plastic ones." I argued that the book is the greatest design of the last two millennia.

But the extraordinary thing about last summer's *Times* story was that we had seen it all before. Xerox had already announced the death of books and the imminent rise of (in effect) "E-books" way back in the 1970s. One thing is for sure. In another decade or two, there will be another *Times* headline: "DESPITE LATEST EXPERT PREDICTIONS, BOOKS SEEM TO BE HOLDING ON." (Intellectuals are people who specialize in recycling other people's mistakes.)

Fifty years from now, we will still be reading books printed on paper and looking at paintings on canvas. If you are lucky enough to listen to Beethoven and watch Fred Astaire movies today, you will still be doing so (if you happen to be alive) fifty years from now.

3) *Tangible gains always trump intangibles.* The book doesn't beat the computer screen for sentimental reasons or because the book is an elegant, aesthetically satisfying

object (although it is). It beats the screen because of its practical advantages. It is inherently more portable, browsable, flip-throughable, write-onable, and readable than anything onscreen. But in general the law of tangibles spells death within fifty years or less for many of today's familiar practices.

Shopping is a classic case. Everyone agrees that small, friendly, "Main Street" stores are better than mall stores, which are better than warehouse stores. No one likes warehouse stores, but lots of people use them. Who could enjoy book shopping at Amazon.com, where you can't flip through the books or even touch them? Yet book buyers abandon bookshops for Amazon because tangible advantages—convenience, selection, sometimes price—trump intangibles every time.

Universities would be obsessed with this law and terrified of its consequences if they weren't too complacent to care. Online education is already emerging all over the landscape. What does a university offer to justify its existence, if you can take all the courses you like online and the quality of online courseware keeps improving every year? Universities are in the business of selling intangibles. They offer the intangible campus experience—putting you face-to-face with your teachers and (more important) your fellow students, and with the campus itself. And therefore 95 percent of the world's universities will be dead in fifty years. The top schools will hang on, because they do sell something tangible—prestige, which translates into jobs and money. But of course they, too, will change. For example: English departments are a luxury good, created originally to teach students great literature. Nowadays many English departments hold that there is no such thing as great literature. Society's response is not (not for long) going to be "Is that so? Then you might as well teach your students any old garbage." It will be "Is that so? Then we don't need an English department anymore, do we?"

Of course, grade schools will bite the dust, too.

4) *The Glenn Gould Memorial Law: Ultimately, technology is a means, not an end.* The great pianist has been dead for almost twenty years. Gould loved technology and mastered it. In the early 1960s, he made the daring prediction that recording would replace live performance. He abandoned the stage and retreated to the recording studio. He cared about every aspect of his recordings—the musical *and* technological details. As twentieth-century masterpieces go, Gould's recordings continue to rank among the best loved. Gould the technology enthusiast embraced technology as the replacement for a centuries-old performing arts tradition. But he used the latest in audio engineering to record . . . the piano. (Or, one time each, the organ and the harpsichord.)

Today we are obsessed with technology (pick up any newspaper). This is an unhealthy obsession; we blather on about technology so that we can avoid topics that make us feel nervous and guilty. If we talk about technology, we don't have to talk about art or science, or truth or beauty, or a parent's moral and spiritual (versus dollars-and-cents) obligations to a child. Instead of moral and spiritual mediocrity, we can talk about financial and engineering brilliance.

Technology is a fascinating topic, and a deep one. But ultimately (as Gould knew) it is a means and not an end. Fifty years from now, technology will be vastly more powerful and ubiquitous than it is today—which is saying a lot. But we will be less and not more fixated on it than we are now.

Certain hardware facts about the next few decades are self-evident. Computer and high-volume memory chips will be so cheap that they will be routinely installed, by the tens of thousands, in the framework of every building that goes up, private or commercial. You'll replace them grudgingly every now and then, like roof tiles.

Fifty years from now (or a lot sooner), the Internet will give way to the Cybersphere, which is full of information beams. When you tap into one of these beams, you tap into an externalized mind—your own or some organization's or institution's. The information you rely on—your own life story plus the equivalent of a few hundred favorite Cybersphere sites—will live wherever you live, and move around with you. You won't carry these information structures; they will move by themselves through the Cybersphere like sociable dolphins following you as you stroll down the pier.

One result will be a huge gain in security. Every tidbit of data (encrypted, of course) will be replicated thousands of times all over the Cybersphere. Every data structure will be distributed—scattered—over thousands of separate micromachines. To destroy or steal data will require a huge number of separate, coordinated break-ins. Your own private encryption key, which you carry in your wallet, won't merely allow data to be decoded; it will allow thousands of separate, individually meaningless strands to be put together into a meaningful big picture—a picture that exists only while you are looking at it.

A thought experiment: Imagine a beam of light cutting through a room from the center of one wall to the center of the opposite wall. You stand in the middle of the room, with the beam passing right in front of you. Of course, light itself is invisible until it strikes something, so the beam is actually a lit-up shaft of dust specks or mist droplets or other small floating particles. An information beam moves "at the speed of time," like a clock: Each lit-up mist droplet (say) moves steadily from the right side of the room (the future) toward the center (the present, the "Now") and on toward the left (into the past). You can "tune in to" the beam—stick a screen (like a tennis racket) into the beam at right angles to the direction of its motion. This is your "beam tuner." The beam sails right through your tuner at a steady rate. By looking at the screen, you can look at the beam as it sails past.

An actual information beam is a stream of information items sailing past. Suppose we convert C-SPAN into an information beam. We might imagine C-SPAN as a stream of freeze-frames, each with a bit of sound attached. (Watching C-SPAN on TV, you see each frame followed momentarily by the next.) Imagine all these frames strung out in a beam. Imagine standing at the center of the room. The right half of this beam is blank; the right half is the future, and those frames haven't been broadcast yet. When C-SPAN broadcasts a new frame, the frame materializes at the center of the room, at the "Now" line, and flows steadily off to your left into the past. If you put your beam tuner at the "Now" line (or infinitesimally to the left of "Now"), you see each new frame as it is created. So you're watching C-SPAN—trapping it (in this thought experiment) on the screen of your beam tuner, frame by frame.

Someone standing to your left can tune in to the same beam. He's also watching C-SPAN, but he's standing (say) ten minutes to the left of "Now" (five feet, say, to your left); he's watching C-SPAN ten minutes in the past, ten minutes behind real time. The C-SPAN beam extends far off to your left. You can watch it an hour, a year, or a decade behind real time, depending where you tune in. What about the right side of the beam, the future? C-SPAN has plans for the future: a programming schedule. It stores this schedule in the "Future" part of the beam. If C-SPAN is broadcasting *Spanorama* at 10:00 a.m. next Tuesday, it puts a note in the beam's future at 10:00 a.m. next Tuesday; the note flows steadily toward "Now." The whole schedule flows second by second toward "Now," where it stops being a schedule and turns second by second into a broadcast.

Information beams are significant because you can store your "information life" in one. And you can store an institution in one.

Take your information life. The beam is a sequence of every information item you create or receive. A beam item can be a picture, a video or audio snippet, a document, a fax, a Web bookmark, or any other unit of information. (The elements of the C-SPAN beam described above are homogeneous; most real beams are wildly heterogeneous.) At the start of your beam, way off to the left and moving steadily farther away, is your electronic birth certificate. New email materializes right in front of you, at the "Now" line. All the documents you've ever worked on are located somewhere in the past, on the left side of the beam, flowing steadily farther away. To work on a new version of some document, you make a copy of it, put the copy at the "Now" line, and work on it there. (Your word processor is corrected for drift: focus it on some document; the document drifts into the past but remains fixed on your screen.) In the future of your stream you store plans, appointments, reminders. They all flow steadily toward "Now," then cross the "Now" line into the past and drift off into history.

I'm leaving a lot out; I haven't explained the hows and whys of this beam. But the point is that this beam captures a documentary history of your life. If you monitor your beam at the "Now" line, you monitor each new piece of information as it arrives. (Phone conversations are part of the beam, too.) Your whole past remains accessible on the beam; your future, as far as you know it, is on the beam, too. (One small but important issue among thousands: Your medical records are stored in your beam. They belong to you; they are directly accessible anywhere to anyone you designate.)

The Cybersphere of the future is full of information beams.

If you work, the "information life" of your company flows through a beam. Each comment, announcement, new order, or family-snapshot-for-general-discussion appears at the "Now" line and flows on back into the past. Everyone has his or her own private view of the company beam. Your own documents and email, visible to you only, are interspersed with group and company items. Discussions take place on the beam and then flow off into history. Instructions are formulated and orders issued on the beam. Factory managers monitor the beam, get orders from the beam, and translate them into action. Company plans, meetings, projections, deadlines are stored in the beam's future.

The beam is the company's mind. Companies have never had "minds" before. A company's mind doesn't act like yours; it is built by many people, not just one. It records the past, present, and future and forgets nothing. By tapping into the company beam, you enter its mind; in fact, you become part of its mind.

A school or university acts in the same way. One way to think of a class or course is as a beam that presents course material point by point, entry by entry. Many students might be working their way upstream simultaneously, each at a different point; meanwhile the teacher watches the entire beam, updating material as needed, monitoring questions as they arise. An electronic campus is a beam. What is its "campus life"? An ongoing discussion (plus physical contact, which you will have to manage on your own). On a campus beam, hundreds of discussions echo forward simultaneously, splitting apart and rejoining; meanwhile the university itself "talks," in announcements, decrees, and so on, each one a separate beam item. All these information units are interspersed with the ongoing discussion. The university posts its plans and schedules in the future, and the campus discussion flows backward into the past. The campus is no longer one point in time; its whole history is on the beam, and you can dip into it anywhere . . . or rather, anywhen. Nor is the campus (or the company) a point in space. You can dip into a beam and join some communal mind wherever you are—sitting in your living room or flat on your back at the beach.

A market is a beam, where buyers and sellers rendezvous.

But for the most part we like to be with other people. We don't want to sit at home. Fifty years from now, people will go someplace and enter a group because they want to, not because they have to. A school will be a random collection of neighborhood children. Each child will tap into a separate beam—the twenty children sitting in a classroom at a "neighborhood school" might actually be attending twenty separate schools, but they can all have lunch and romp outside together; any responsible adult (with or without an ed school degree) can keep an eye on them. "Neighborhood offices" will work the same way. You might work in a small office complex with a couple of dozen people; they might all work for different companies but still choose to spend their workdays together.

The institutional office buildings that shape our landscape today will disappear. Stores and shops are already on the way out. (E-commerce seems to be hung up temporarily, but remember how absurdly primitive today's Web sites are. There is no reason, for example, why you shouldn't be able to flip through the books at a bookstore site. There is no reason why you should have to master a new layout every time you visit a new site. For better or worse—and the consequences will be distinctly a mixed bag—commerce and education are moving inexorably into cyberspace.) The end effect of world-spanning information beams will be to make neighborhoods as important as they were in the nineteenth century. People will need houses and convenient, generic, local public gathering spaces. We won't need cities anymore, except as gigantic museums/theme parks/shopping malls. Which is too bad—cities are among humanity's greatest artworks. But we are bound to appreciate them more when we no longer need them.

Will people drive less fifty years from now? No, they will drive more. Figure it out: We *like* to drive. The richer we are, the more we can do what we like.

So the world half a century from now will look different and work differently. It will be much richer. It will have far snazzier technology. It might even be ever so slightly happier.

❑

DAVID GELERNTER is a professor of computer science at Yale and chief scientist at Mirror Worlds Technologies (New Haven). His research centers on information management, parallel programming, and artificial intelligence. The "tuple spaces" introduced in Nicholas Carriero and Gelernter's Linda system (1983) are the basis of many computer communication systems worldwide. Dr. Gelernter is the author of *Mirror Worlds, The Muse in the Machine, 1939, Drawing Life*, and *Machine Beauty*.

JOSEPH LEDOUX

❑

Mind, Brain, and Self

YOUNG SIGMUND FREUD BEGAN his scientific career by studying the nervous system and believed that the secrets of mental life would be illuminated by an understanding of brain function. Because he soon came to realize that the tools available for studying the brain were not sufficiently advanced to put his belief into practice, he turned to a purely psychological approach. In the intervening years, neuroscience has become a thriving discipline, and its discoveries would astound Freud. Still, there is much left to learn, and some of the developments we can expect in the coming years are described here.

Reading the Brain

Neuroscientific research has gone a long way toward revealing how certain aspects of the mind, like perception, memory, and emotion, are mediated by the brain. Much of this work has involved studies of nonhuman organisms, especially rats and monkeys. While this approach is adequate for asking questions about brain functions that humans share with other creatures, it has left important gaps in our understanding of the unique aspects of the human brain. Research on humans with brain damage has helped close the gap, but studies of brain-injured subjects are as much about how the brain compensates for lost function as about normal function.

New technologies are enabling us to study normal human brain function, and they promise a new level of under-standing of the relation of the human brain to the human mind. Specifically, the emergence of functional magnetic resonance imaging (fMRI) has provided a safe and practi-cal means for researchers to peek into the brain of a human being and observe its activity as the subject performs psy-chological tasks or has certain experiences. Most of the imaging studies to date have focused on validating the technique, showing that this approach reveals the same picture of brain function as more traditional approaches. Many of the existing findings are thus referenced to studies of brain function in experimental animals. Without this basic grounding in how specific brain systems work, the imaging findings would exist in an intellectual vacuum. For example, studies of rats and other mammals have shown that the amygdala, a small region in the temporal lobe, is a key part of the brain network involved in detecting and responding to danger. Guided by this information, researchers then showed that patients with amygdala dam-age are impaired in recognizing danger and that regions of the amygdala are activated, as determined by fMRI, when humans are exposed to threatening stimuli. In this and many other areas, animal studies have paved the way.

It is important to match the species studied to the ques-tion being asked. For example, working memory, which allows you to hold information in mind and do things with it, is a key process underlying human thought. This process is known to involve a region of the human brain called the dorsolateral prefrontal cortex. Rats do not have a dorsolateral prefrontal cortex and thus are not appropriate subjects for these kinds of memory studies. Monkeys do have a dorsolateral prefrontal cortex, and much of what has been learned about the role of this region in working memory was discovered through studies of monkeys. But a key aspect of human thought involves verbal working memory, a function that cannot be studied directly in any species but humans. Recent fMRI studies have played an important role in elucidating how verbal working memory functions in the human brain.

The future of research on the human brain with fMRI or other approaches—including other ways to record activity, and ways to stimulate selective brain regions and induce activity—is likely to be in three broad domains. The first is the most pedestrian: We will learn more about some processes that we already know something about—that is, the neural organization of perception, memory, emotion, language, and working memory. The second entails discovering more about how these processes interact in the brain. This investigation will take us from narrow to broader systems-level concepts of brain function, and toward at least the beginnings of a theory of how the brain makes the mind, as opposed to how specific mental processes function. Work of this type has begun, but is far too rare.

The third domain is perhaps the most important. Nearly all studies of brain function focus on the way the brain typically works in most of us most of the time. Each such study includes a great many subjects, in order to produce a norm. Once we have a solid grounding in our understanding of these normative functions, we can ask questions about how variations between individuals determine the unique qualities that account for the self or personality. These questions require a slightly different approach—one in which a lot of measurements are made in an individual, rather than a single measurement being made in multiple subjects.

Existing techniques are giving us powerful tools for assessing what is going on in the brains, and minds, of people. As these techniques improve, we will have to ask whether we are as a society ready for what this research will tell us. When it becomes possible to look inside the brain and see what someone is thinking or feeling—to predict, say, whether someone is likely to be a murderer, child molester, or rapist—what will we do with this information?

Managing Memory

Every time you form a memory, you adjust the wiring—the synaptic connectivity—of your brain. Be it as trivial as the color of the socks you pulled up this morning or as significant as the sound of your mother's voice, memory is a process of adjusting connections between neurons. In simple terms, it goes like this: Those neurons that are actively engaged during an experience undergo certain chemical changes that activate genes and thereby initiate the synthesis of proteins inside these active cells. The proteins are then shipped to the active synapses on the active cells, where they alter the ability of those synapses, and only those, to receive messages from the neurons they are connected with. Memory is embodied in such changes. We can expect, given what we already know, that in the near future it will be possible to manage memory in various ways.

Now that people are living longer, more of them are suffering from age-related memory problems. These problems are most apparent in people with Alzheimer's disease and certain other neurological conditions, but memory also falters in older people without specific brain disorders. Scientists are currently attempting to use information gleaned from memory studies of such diverse animals as sea slugs, flies, rats, rabbits, and monkeys to develop ways to improve our human memory. It is well established, for example, that many forms of memory depend on the neurotransmitter glutamate and its receptors. One strategy for memory improvement thus involves the development of drugs that facilitate glutamate transmission. Further, an important step in memory formation is the flow of chemical ions (especially calcium) through glutamate receptors into neurons; the rise in calcium then leads to the activation of molecules that in turn activate genes. The development of drugs targeting these processes within our brain cells—that is, attempting to improve their ability to activate the genes that make the proteins that stabilize the synaptic wiring that underlies memory—offers another strategy for improving memory function.

But what about fixing the brains of people with neurological problems like Alzheimer's disease? The recent discovery that in adult brains new neurons are made in the hippocampus, a brain region of central importance to our ability to consciously remember, offers new hope. If these cells can somehow be encouraged to connect with and thus participate in the degenerating memory circuits, perhaps memory function can be restored. And if the federal government will untie the hands of researchers and allow them to proceed more freely with stem cell research, it may be possible to prevent conditions such as Alzheimer's from being expressed at all in people who are susceptible.

Another area where brain science could have an important impact is in the prevention or elimination of unwanted memories, especially traumatic ones. Such memories form the core of conditions like posttraumatic stress disorder, and if they can be short-circuited, the disorder might be ameliorated to some extent. Researchers have come up with ways to alter the fate of memories while they are being formed and stabilized; this could lead to the development of drugs that could be administered shortly after some highly stressful event and thus prevent the development of traumatic memories. But because the stabilization of memory formation takes only a few hours—the time it takes for proteins to be made and utilized—this approach will have limited application. An alternative, though, may be available.

New studies in rats have shown that specific well-formed memories can be disrupted if proteins are interfered with at the site of memory in the brain during the process of recalling the experience. But in order to be useful in dismantling traumatic memories in humans while leaving other memories intact, the operative drug would have to target the areas involved in traumatic memory. This in turn will require us to find the site of traumatic-memory formation in posttraumatic stress disorder, as well as some way to restrict the drug to that region. We'll consider these points shortly.

Of course, even if it becomes possible to weaken or remove disturbing memories in humans, it is not something we should do lightly. Imagine a Holocaust victim who lived for decades with memories of death camps. These memories have undoubtedly become ingrained as part of the victim's personality. Although the person may be severely troubled by such memories, what would happen to the fabric of her personality if a set of episodes that had become such a central part of her life were removed?

Scientific advances sometimes become part of daily life. We might therefore see the day when over-the-counter drugs will be used to give a particular experience an especially strong representation in your brain. Suppose you want to remember a birthday or wedding anniversary particularly vividly. Right before the party, pop a pill that gets glutamate or other molecules working more efficiently, and everything that happens will be burned into your circuits in brilliant detail.

Recreational rewiring is not as far-fetched as it sounds. We arrange situations all the time to increase the emotional impact of experiences and make our recollections of them vivid and enduring. Taking a drug to do this is just a different way of doing the same thing. It is less romantic to give your spouse a pill on your anniversary than a bouquet, but the pill may achieve the desired result (a memorable evening) more effectively. Or you can hedge your bets and try both the pill and the bouquet.

Smart Drugs

Macbeth pined for "some sweet oblivious antidote" to sorrow. We now have a number of drugs that are fairly successful at helping to treat depression and other psychiatric disorders. But drugs come with a price—side effects. Fifty years from now, or sooner, drugs will treat troubled networks in the brain without affecting others. To create such drugs, several developments will have to take place.

We will first need to learn more about exactly which networks are troubled in specific disorders. Brain imaging is already beginning to be helpful in this respect. Studies are showing how the brains of people with depression, anxiety disorders, or schizophrenia differ from those of people without such afflictions. But in order make some sense of these differences, we need to learn more about the normal function of the areas identified.

For example, it is a reasonable assumption, given existing animal and human data, that fear-related disorders (panic attack, posttraumatic stress disorder, generalized anxiety, phobia, paranoid schizophrenia) result from alterations in the way the brain's fear networks normally function and interact with other networks. Since the amygdala, as we have seen, is a key part of these networks, alterations in amygdala function might account for certain aspects of anxiety. Specifically, excess and inappropriate fear could occur because the amygdala is oversensitive, detecting danger and responding defensively to a situation that would be ignored by another person; or the amygdala could be too reactive, responding with a more vigorous defense than another person would to the same degree of threat. Either of these conditions could arise from genetic wiring or from traumatic or otherwise stressful experiences, or from some combination of the two. Moreover, either effect can be accentuated by the way other brain regions connected with the amygdala regulate amygdala function. And different conditions may be accounted for by different alterations of circuits within the amygdala, or between the amygdala and other areas. If imaging studies determine that the amygdala (or any other area) is affected in anxiety disorders, clarification of the function of the region and its interaction with other systems will be fundamental to inventing new treatment strategies. But even now, when imaging studies are showing the involvement of certain brain regions in such human conditions as anxiety, animal studies remain important for understanding detailed neural mechanisms at the level of cells and synapses in that region; ultimately, the development of new and better medications depends on this level of knowledge.

Once human imaging studies implicate specific networks in a particular condition and animal studies illuminate the detailed organization of those networks, we can look for drugs that will target the afflicted circuits. One strategy would involve capitalizing on advances in molecular genetics: If we can identify some molecule that is expressed only in the amygdala, or is expressed there in some particular way, it might then be possible to use that molecule as a key to unlock a drug. That is, the drug would still be taken orally and would still travel widely in the bloodstream to many brain regions; however, because of the drug's molecular packaging, it would be inert in most brain regions. Only when it encountered the molecular key, which in this hypothetical example is present only in the amygdala, would the drug become active. Such a drug could help correct abnormal amygdala function without affecting other brain regions, thus reducing unwanted psychological side effects caused by widespread drug action. But because the amygdala also participates in "normal" brain functions, the real challenge will be to find some way to selectively attack the disordered functions.

The Amygdala Defense

The amygdala, like many brain regions, does its work outside our conscious awareness. We can become aware of the consequences of amygdala activation, but we do not have conscious access to its inner workings. Because the amygdala can be provoked into expressing unconsciously controlled emotional responses, the possibility is raised that the amygdala could unconsciously commit a crime—one that the conscious person would never willfully condone.

This possibility has not escaped lawyers. The legal system has long recognized "crimes of passion," in which an otherwise law-abiding and reasonable person commits a crime during a lapse of rationality or sanity. The "amygdala defense" adds a neurological rationale to this sort of argument. As we learn more about how the brain works, and lawyers learn more about what has been discovered, neurologically based defenses will become more and more common. So let's take a close look at what I mean by the amygdala defense.

First, the amygdala defense should not be confused with a related issue, which we can call the pathological brain defense. In the latter, the argument is that the person committed a crime because of some physical alteration in his or her brain. The amygdala defense, in contrast, is based on the notion that the amygdala normally controls emotional behavior in an unconscious fashion, and as a result it is possible for a crime to be committed by the amygdala independent of conscious thought. It is clearly possible for the amygdala to control an aggressive act independent of conscious control in certain provocative circumstances; however, in order for the amygdala defense to work, several criteria would have to be met.

An important job of the amygdala is to rapidly initiate protective responses in the face of a sudden danger. But if the stimulus has been present for some time and consciously perceived, behavior tends to be under the control of higher thought processes, mediated by the cortex. Further, the kinds of responses directed by the amygdala are fast, simple, innate (hardwired) responses that are exe-

cuted in a stereotyped manner—that is, performed similarly in all members of the species. So if an act is deliberate,
expressed relatively slowly (over seconds rather than mil-
liseconds), involves a complex sequence of movements,
and would be carried out differently in different people, it
is probably not directly controlled by the amygdala. The
amygdala can indirectly influence or modulate these more
complex responses, but they are, in the end, the business of
other brain systems. These facts suggest that in order for
the amygdala defense to succeed, the crime would have to
involve a relatively simple, innate, stereotyped response
executed instantaneously and without premeditation upon
the occurrence of the provocation.

I suspect that few crimes would meet the criteria necessary for the amygdala defense to succeed. However, it is
becoming increasingly apparent that many brain systems
other than the amygdala function unconsciously—and even
that consciousness itself is the product of the unconscious
workings of brain networks, raising the possibility that
while the amygdala defense is wrong in name, it may still
be valid in spirit. Whether we will need to reconsider the
nature and limits of human responsibility, though, will
depend on future discoveries about the balance between
conscious and unconscious control in the brain. These, too,
are likely to come in the next fifty years.

❑

JOSEPH LEDOUX is the Henry and Lucy Moses Professor of
Science in the Center for Neural Science, New York University. He has long sought to understand our emotions as
biological states of the brain. His work emphasizes the role
of learning and memory (in contrast to genetic predetermi-
nation) in emotional experience and seeks to relate memo-

ries of emotional experiences to synaptic events. His newest book is *Synaptic Self: How Our Brains Become Who We Are*. He is also the author of *The Emotional Brain: The Mysterious Underpinnings of Emotional Life*; coauthor (with Michael Gazzaniga) of *The Integrated Mind*; and editor (with W. Hirst) of *Mind and Brain: Dialogues in Cognitive Neuroscience*.

JUDITH RICH HARRIS

❑

What Makes Us the Way We Are:
The View from 2050

As THE OLDEST LIVING MEMBER of the Society for Research
in Child Development (I turned a hundred and twelve in
February), I've been asked to report on scientific advances
made in our field over the past fifty years—the first half of
the twenty-first century. Before I turn to children, however, I would like to say a word about old people like me.
Aging, as the members of this organization are aware, is
also a kind of development. Among the most important scientific advances of this century, in my view, was the introduction of drugs that can prevent, and even to a certain
extent reverse, the changes in the brain associated with
Alzheimer's disease. Though I don't pretend that my memory is perfect—I ask your forgiveness if I fail to mention
something you feel should have been included in this
report—the very fact that I was invited to come here today
and deliver it to you is a testimony to the efficacy of those
drugs.

That said, I'll turn to my assigned topic: advances in the
field of child development in the past fifty years. When the
twenty-first century began, developmentalists had already
learned a lot about those aspects of development that are

basically the same for all normal children. Good progress had been made in understanding how children learn to think, to talk, to read, and so on. But very little was known about what made them grow up different from one another—why one child would become a kind, conscientious adult, another an aggressive or impulsive one. Developmentalists of the twentieth century (they called themselves "developmental psychologists" back then) thought they understood the sources of individual differences in behavior and personality, but as we now know, they were mostly wrong. Thus, the most important advances made in the twenty-first century have been in understanding why people differ from one another and in putting that understanding to use.

Before I describe these advances, I think it's informative to look at the reasons that so little progress was made in this area in the previous century. The main reasons were a disdain for genetics and the use of outmoded research techniques. By the year 2000, developmentalists had grudgingly admitted that babies are not all alike at birth—that each is born with distinctive characteristics, largely genetic in origin. But they were still using a research method that had originated in the 1950s and that was based on the assumption that babies start out all alike!

You see, in the 1950s most developmentalists really did believe that newborns are all alike and that any differences found among them at a later age must therefore be due to differences in what they experienced after they were born—differences in environment. The research method that originated at that time made sense, given that assumption, but unfortunately it continued to be used long after the assumption was discarded.

The method was simple. Developmentalists in the area called "socialization research" would measure some aspect of the environment and some aspect of the children's development. Then they would look for correlations between the environmental measure and the developmental measure. Next, they would report their findings: for example, that children whose parents read to them a lot tended to become better readers, or that children whose parents spanked them a lot tended to be more aggressive, or that adolescents whose parents had heart-to-heart talks with them were less likely to get into various sorts of teenage trouble. The final step was to turn these findings into recommendations to parents: Read to your children if you want them to do well in school. Refrain from spanking them if you don't want them to be aggressive. Have frequent heart-to-heart talks with them if you don't want them to get into trouble. The government of the United States actually paid researchers large sums of money to carry out this kind of research and to issue these recommendations!

Yes, we can laugh about it today, but it was a serious matter back then. The developmentalists of the late 1900s were "in denial," to use a phrase that was popular at the time. They hadn't faced the fact that if genes made any significant contribution to the outcomes they were measuring, the results of their research were uninterpretable. Though they'd admitted that children have genes, they hadn't yet admitted that children inherit their genes from their parents and therefore tend to resemble their biological parents in many ways—in intelligence, aggressiveness, and conscientiousness, for instance—for genetic reasons alone.

Why did it take the developmentalists so long to

acknowledge this obvious fact? After all, studies using better techniques had already produced enough data to show that they were jumping to the wrong conclusions. Researchers in the field then called behavioral genetics (now better known by the names of its subdisciplines) had shown that the correlations the twentieth-century developmentalists were so fond of reporting could be explained almost entirely on the basis of genetic similarities among members of biological families. The correlations disappeared if you looked at adoptive families. But these results, and the admonitions written by people who understood them, were largely ignored.

I was one of those people who, around the turn of the century, wasted my breath issuing admonitions. Someone predicted that my 1998 book *The Nurture Assumption* would become "a turning point in the history of psychology," but alas it was not to be. Such a big ship cannot turn on a dime. It was steaming full speed ahead, steered by academicians who were highly thought of at the time and who were quite comfortable with the status quo. Many nudges were required to turn it in a new direction. The first, if I remember correctly, was a book that appeared a few years before mine: *The Limits of Family Influence*, by David Rowe. Shortly after the turn of the century came Steven Pinker's *The Blank Slate*. A few years later, there was a book by Robert Plomin. Still later came the contributions of Eldrick Woods and Abigail Valk. (As a historical aside, some of you might not know that Woods had an earlier career as a champion golfer. Abigail, of course, is not only a past president of this organization but is also my granddaughter.)

But the biggest nudge came from outside the field of developmental psychology—sorry, I mean developmental

science. The decoding of the human genome gave tremendous impetus to research in genetics, and this led first to an
appreciation and later to an understanding of how tiny differences in genes could produce noticeable differences in
people's personalities and cognitive abilities. Researchers
finally faced the fact that they couldn't figure out how
the *environment* affects a child's development unless they
knew what characteristics and predispositions the child
had brought to that environment. Without controlling for
genes, research that looked only at outcomes of development can't tell us anything.

Now we can control for many kinds of genetic effects
directly, by scanning a person's genome in search of various
combinations of genes. But for a long time, controlling for
genetic effects required the use of more laborious methods, such as studying adoptees or twins. Though these
methods were clearly more productive, the older methods
continued to be used until 2016, when the United States
government finally put its foot down and refused to fund
any more developmental research that didn't include
genetic controls. The decision revolutionized the field—
not only because it put an end to useless research but also
because so many of the older generation of developmentalists decided to retire at that time.

There were, of course, other factors that helped turn our
field in new directions. I'll mention one other: the knowledge gained from discoveries in paleoanthropology. The
most important one was made in 2021, in a melting glacier
in . . . oh, where was it?—somewhere in Scandinavia or
thereabouts. Found in the ice was the body of an early
European who had died about twenty-seven thousand
years ago. But it wasn't the European himself that caused
the commotion; it was what he was wearing. His coat was

made out of a beautiful, thick fur that no one could identify at first. Well, as you know, the fur turned out to be that
of a Neanderthal—of three Neanderthals, to be precise.
The discovery led to a dramatic revision of ideas about the
evolution and history of the hominids. It demonstrated
something the paleoanthropologists should have realized
all along: Neanderthals were furry. They *couldn't* have survived for so long in Ice Age Europe without a heavy coat
of fur; a deer hide slung around the shoulders simply isn't
adequate protection against that kind of weather, and they
hadn't invented the needle, so they couldn't sew. It was
hard at first for people to accept the idea that our ancestors
saw Neanderthals not only as a source of food (that seemed
excusable, since the Neanderthals had the same attitude
toward us) but also as a source of clothing.

Though that discovery only proved something we already
knew—that we are predators, the most deadly ones the
world has ever known—it ultimately led to a more realistic
view of human nature. The romantic notion of the "noble
savage" was finally consigned to the dustbin; our ancestors
were savages, all right, but they weren't noble. We didn't
get where we are today by being nice guys.

Clearly, though, we are capable of being nice guys under
the right conditions. Researchers have made good progress
in specifying those conditions, but there is much work yet
to be done.

What We've Learned about Child Development

I apologize for talking so much about genetics when most
of us here are more interested in the effects of the child's
environment. But, as I said, in order to see environmental

effects clearly, we first have to skim off the effects of the genes. With twenty-first-century technology and methodology, researchers can now do that fairly accurately.

This research has shown that though the environment does have important effects on children's development, it doesn't work the way the twentieth-century developmentalists thought it did. Most of the correlations on which they based their theories turned out either to be direct effects of genes—the fact that parents and their biological children have similar genomic profiles—or to result from the way parents react to their children's behavior. Parents are less likely to have heart-to-heart talks, for example, with teenagers who are contemptuous of everything they say or who refuse to listen to them. As it happens, these are the very teenagers who are most likely to get into trouble. The failure of the parent to have heart-to-heart talks and the tendency of the teenager to do foolish things are correlated because they have the same source: the teenager's personality.

So the question is, How do we account for the teenager's personality? Why is one sensible and another impulsive, one nice and another nasty? Decades before the beginning of this century, we knew it wasn't all genes; genetic variation could explain only about half the variation in personality from one individual to the next. But little was known about the nongenetic influences on personality development, because most of the research time and money had been wasted on what proved to be dead ends.

A breakthrough occurred when developmentalists finally came to appreciate the interplay between personality and context. It had long been recognized that personality is strongly influenced by context—that people behave differently in different contexts—but that there's some carryover of behavior from one context to another. Twentieth-century developmentalists misinterpreted the carryover. They saw that some children were troublesome both at home and in school, for example, and jumped to the conclusion that the child's troublesomeness at school was due to something that happened at home. But once we had the methodology to separate genetic and environmental influences on behavior, it became clear that any tendency for children to behave similarly in different contexts is due almost entirely to genetic influences on their behavior. The environmental influences don't transfer from one situation to another (though if the situations are similar, the environmental influences in the two situations will also be similar, of course).

That understanding cleared up the mystery of why the childhood home environment appeared to have so little effect on the way children turn out. What happens within the family does matter, but in order to see its effects you have to observe how people behave with their parents and siblings. In adulthood, personality is seldom assessed under those conditions, so naturally it doesn't show effects of the childhood home. The reason that parents don't have long-term effects on their child's personality is simple: People don't spend their adult lives in their parents' home.

Thus, to find out how childhood environment affects adult personality, researchers had to focus on what happens to children outside the home. Almost everything that happens to children outside the home turned out to make a difference—their experiences at school and in the neighborhood, the way they are treated by their teachers and their peers. We already knew that the culture is important, but we learned that the culture has long-term effects only if it's transmitted by some agency other than the parents. If it's transmitted solely by the parents, children assume that it is idiosyncratic to their home and family and inappropriate or irrelevant elsewhere. Parents realized that if they wanted to transmit a particular culture to their children, they'd have to rear their children in a place where the children would be exposed to that culture outside the home as well as within it. Most parents were doing that anyway, but it was interesting to find out why it worked.

We still have much to learn. At the turn of the century, I optimistically thought that by 2050 we would have found the source of most of the nongenetic variation in personality (the variation attributed to what was then called the "nonshared environment"). But so far we've managed to account for only about half of it, which leaves about a quarter of the variation in personality still unexplained. We know that some of the unexplained variation is environmental—little things that happen to children along the way which are hard to study and even harder to predict. But some of the variation has a biological explanation— nongenetic, yet biological. Even babies with exactly the same genes (identical twins or clones) are not precisely the same at birth; just as their fingerprints differ slightly, so do their brains. Researchers are currently studying the molecular processes responsible for these subtle differences in brain formation and the effects these differences have on personality, but this work is still in its infancy.

So we still can't predict behavior or personality with anything close to certainty. People tend to find this lack of certainty annoying when it comes to predicting the behavior of other people, but for some reason they like the fact that other people can't predict *their* behavior!

Making Use of What We've Learned

We know now that what happens to children at home affects their behavior at home, and what happens to them outside the home affects their behavior outside the home. If a child's behavior is causing problems at home, we can help the parents by teaching them more effective child-rearing skills. If a child's behavior is causing problems at school, it's the school's responsibility to deal with it, and we can help there, too. We've learned, for example, how to keep smaller, weaker, or less attractive children from becoming the victims of bullies.

We've learned that children need a stable environment outside the home and that it's bad for them to be moved around too often—to be shifted repeatedly from one neighborhood or school to another. They need, in particular, a stable peer group, so that they don't have to repeatedly win acceptance from a new group of peers and repeatedly adapt themselves to the new group's standards of behavior, dress, and speech. We've learned that it doesn't much matter how many parents children have (much less what gender they are), as long as changes in the parenting arrangement don't disrupt the child's life outside the family. Just as people in the mid-1900s used to frown on parents who split up the family because they decided they didn't want to live with each other anymore, people in the mid-2000s frown at parents who move their child around for their own convenience.

Fortunately, nowadays people generally put off having a child until they are able to provide it with a stable home. Advances in reproductive medicine have pretty much put an end to accidental pregnancies. The downside of our ability to control conception is that we don't have very many children. Though most governments are doing what they can to encourage reproduction, the population is declining in nearly every part of the world.

From the child's point of view, this is all to the good. The competition for teaching jobs means that all our teachers are highly qualified (and highly paid). Children are taught in smaller classes and smaller schools, which has turned out to have benefits that go far beyond the obvious educational ones. Teenagers used to attend very large high schools and as a result they tended to split up into adversarial groups: proeducation vs. antieducation, athletic vs. nonathletic, brown-skinned vs. pink-skinned. The effects were often quite harmful. With smaller classes and smaller schools, that is less likely to happen. If it does happen, we've learned what to do about it.

The knowledge we've gained has had tremendous benefits for parents. Parenting in the second half of the twentieth century was a more difficult job than at any time before or since, because the "experts" had made people feel that their children had fragile psyches and one wrong move might do permanent harm. Parents were afraid to exert their authority. Physical punishment was seldom used; instead they had something called a time-out, which was harder for the parent to administer than for the child to endure. Children were deluged with hugs, kisses, gifts, praise, and declarations of love. The fact that children want endless attention and praise was misinterpreted to mean that children *need* endless attention and praise. In those days, people made a big deal about being "natural," and yet their child-rearing style required them to go against their natural inclinations and to simulate affection they often didn't feel. They had to put aside their own desires—even their need for sleep.

Child-rearing styles undergo periodic shifts, from strict to indulgent and back again. I've lived long enough to see the pendulum swing both ways. The way people reared their children fifty years ago seems ludicrous to us now. Today's children don't get as much verbal and physical affection, but what they do get is genuine. Outside the home, children associate day after day, year after year, with the same small group of peers.

The result, oddly enough, is that in many ways childhood today bears a closer resemblance to childhood in an ancient tribal or hunter-gatherer society than to childhood in a typical American home at the end of the last century. I can't prove a causal connection, but the rate of childhood depression, which hit an all-time high in the early 2000s, is noticeably lower today.

On that encouraging note, I conclude my report. Thank you.

❑

JUDITH RICH HARRIS is a writer and developmental psychologist. A former writer of textbooks on child development, she realized one day that much of what she had been telling her readers was wrong. She stopped writing textbooks and instead wrote an article proposing a new theory of development; her article, published in the *Psychological Review*, received the George A. Miller Award from the American Psychological Association in 1998. Harris's book *The Nurture Assumption: Why Children Turn Out the Way They Do* was a finalist for the Pulitzer Prize in 1999.

❏

Drugs, DNA, and the
Analyst's Couch

IN 1950 A CHEMIST AT Rhône-Poulenc, a French pharmaceutical company, modified the structure of an antihistamine and accidentally created a drug that can eliminate the psychotic thinking of people with schizophrenia. Within a few years the new drug became world famous as chlorpromazine (Thorazine), the first truly effective medication for a disabling mental disorder. Because of its dramatic effect, chlorpromazine set a new course for psychiatry for the rest of the twentieth century.

The great success of chlorpromazine stimulated vigorous competition from other pharmaceutical companies. In the 1950s the search for more antipsychotic medications led to the accidental discovery of two other types of psychiatric drugs. First Geigy, a Swiss pharmaceutical company, came up with a modified version of one of its antihistamines that, although useless against psychosis, proved to be a valuable treatment for severe depression. Named imipramine (Tofranil), it paved the way for many contemporary antidepressants. Then Hoffman-La Roche, another Swiss company, created chlordiazepoxide (Librium), which doesn't help psychosis either but does relieve anxiety. It was soon followed by another benzodiazepine, diazepam (Valium), which became the best-selling drug in America for about a decade, beginning in the mid-1960s.

Adding to the excitement over these drugs were a flurry of findings about their effects on neurotransmitters, a class of brain chemicals that transmit signals between nerve cells. By the 1970s it was discovered that chlorpromazine blocks certain actions of a neurotransmitter called dopamine; imipramine augments the actions of several neurotransmitters, including norepinephrine and serotonin; and diazepam amplifies the effects of yet another neurotransmitter, called gamma-aminobutyric acid (GABA). In each case, the net result is a change in signaling in brain circuits that control emotional aspects of behavior.

These discoveries spurred a search for other chemicals that would have similar effects on neurotransmission but fewer undesirable side effects than the originals. The search paid off in a stream of new medications that patients prefer. The most famous, fluoxetine (Prozac), was initially identified as a chemical that prolongs neurotransmission by serotonin; it was subsequently shown to be an effective treatment for both severe and moderate depression. Called an SSRI (selective serotonin reuptake inhibitor), it prolongs serotonin's effects by inhibiting its reuptake by the nerves that release it, which is the normal way that serotonin signaling is terminated. Related drugs, including sertraline (Zoloft), paroxetine (Paxil), fluvoxamine (Luvox), and citalopram (Celexa), soon followed.

As experience with SSRIs grew, psychiatrists became aware that these medications could also help people who aren't depressed. SSRIs have now become an established treatment for unprovoked attacks of panic (panic disorder) and uncontrollable worrying (generalized anxiety disorder)—beneficial effects that have been confirmed by formal comparisons with placebos in controlled trials.

The effectiveness of these and other new medications transformed psychiatry. Before such drugs came on the scene, most psychiatrists thought about their patients in purely psychological terms and were mainly interested in treating them with psychotherapy. Now attention has shifted to the brain, and psychiatric treatment frequently includes at least one medication. Tens of millions of Americans take psychiatric drugs.

But, valuable though they are, the drugs that replaced chlorpromazine, imipramine, and chlordiazepoxide are simply modified versions of the originals. None of them is substantially more effective, and all of them have some undesirable side effects. Despite extensive knowledge about their effects on neurotransmission, the development of new drugs still relies on a trial-and-error approach similar to that used in the 1950s.

The next big step in psychiatry is not likely to come from further refinements of the drugs and psychotherapies that define the field today. It will come, instead, from discoveries about human genetic variations and the ways they affect the brain. Just as eye-opening stories from psychoanalysts' couches guided psychiatry in the first half of the twentieth century, and the products of smelly chemistry laboratories guided it in the second half, so will knowledge about individual genetic differences guide psychiatry over the next fifty years.

Knowledge about individual genetic differences holds out such promise for psychiatry because it will help to answer a fundamental question: What determines individual susceptibility to disturbed behavior? A person's past experiences clearly play a vital role. But why does one person transcend repeated mental hardships, whereas another is readily tipped into a state of distress? And why does one person succumb by lapsing into depression, another into sustained anxiety, and a third into the withdrawal and delusions of schizophrenia?

The best clue we have is that all these patterns of disturbed behavior run in families. Consider, for example, the risk of becoming schizophrenic. Most people have about one chance in a hundred of developing the characteristic pattern of symptoms. But if you have a parent or sibling who is schizophrenic, your lifetime risk is eight times greater. The same is true of the other major cause of psychosis—manic-depressive illness, also known as bipolar disorder. Again, the general risk is about one in a hundred, but the risk is eight times greater if you have a parent or sibling who suffers from this disorder. Depression and the anxiety disorders are also familial.

Not long ago these studies of families sparked explosive debates between those who took them as evidence for learned familial patterns of abnormal behavior and those who took them as evidence for inheritance of a predisposition to mental disorders. Now most people agree that environment and heredity both play some part. They also agree that the best next step in assessing the importance of heredity is to try to find the alternative forms of genes that are involved.

The main catalyst for this agreement has been the development of powerful techniques for direct examination of the alternative forms of human genes, called alleles, or gene variants. These variants, which arose by random changes in DNA structure, are responsible for a great deal of human diversity, including differences in susceptibility to particular illnesses. But until recently their existence could only be inferred. The new techniques make it possible to identify specific gene variants that contribute to a human attribute. Instead of arguing about the relative importance of nature and nurture, we can now turn our attention to a search for gene variants that contribute to individual predisposition to an illness.

One way to find them is to compare the DNA of family members who have that illness with those who don't. If only those with the illness have a certain variant of a particular gene, the correlation is probably meaningful. If the same variant is also found only in the affected members of a number of other families, the case is strengthened. At some point the likelihood becomes so high that a role for the variant is established. As the details of human genetic structure were being worked out in the 1990s, some gene variants affecting susceptibility to particular illnesses were identified in just this way. Famous examples include the variants of three different genes that each greatly increase the risk of developing rare types of Alzheimer's disease which begin before the age of fifty. In one group of families the culprit was a variant of a gene called APP; in another it was PS1; and in a third it was PS2.

The discovery of gene variants that increase the risk of rare types of Alzheimer's disease has stimulated genetic studies of schizophrenia, depression, manic depression, and other psychiatric disorders. The immense appeal of these studies is that they are not dependent on guesses about which genes might be involved, because they can be designed to detect a correlation between the disorder and a variant of any human gene. Although many early studies did focus on specific genes, especially those that influence neurotransmission, we know so little about the genetic control of mental processes that it would not be surprising if other types of genes were implicated. Unfortunately, despite years of effort, no one has yet found a gene variant that definitely increases the risk of any of these mental illnesses. Nor has there been much success in genetic studies of other common disorders, such as diabetes and high blood pressure. One reason for this lack of progress is that susceptibility to all these maladies is determined by the combined actions of variants of multiple genes rather than by variants of a single gene. Although current technology has made it relatively simple to identify the rare variants of single genes that do indeed have a major effect on risk— such as APP, PS1, or PS2—it remains very difficult to find those gene variants that increase risk only if they are inherited in concert with a number of others.

This difficulty will soon be lessened because of the continued growth of knowledge about the human genome. The recent publication of the detailed structure of human DNA is a critical first step. Now DNA specimens from many people are being examined in order to identify and catalog the common variants of each of the approximately thirty thousand human genes. This will greatly simplify the search for the many gene variants that may operate together to influence vulnerability to mental disorders. The search is also being simplified by the development of efficient new techniques for detailed examination of the DNA of any individual. These techniques are in a continual state of improvement, reminiscent of the ongoing development of computer chips. So, too, are the computational methods used to analyze the masses of information from such DNA studies.

With the evolution of the technology for collecting and evaluating large masses of DNA data, it will soon be possible to mount a massive search for the groups of gene variants that influence susceptibility to particular mental disorders. As the costs of DNA analysis keep falling, we can go beyond relatively small family studies and scrutinize DNA samples from thousands of unrelated people with a particular disorder. Such an investigation should identify the relevant gene variants, only some of which will be found in each affected individual.

To properly use this mass of data about gene variants, it will be necessary to correlate it not only with patterns of disordered behavior but also with properties of the brain. A variety of new methods, such as functional magnetic resonance imaging, are beginning to be used to assess the functions of specific regions of individual human brains. Correlating patterns of gene variants with the results of these and other studies will lead to the identification of subtypes of disorders that are presently lumped together in diagnostic categories, such as schizophrenia or depression.

The combination of genetic information and functional studies will also provide targets for truly novel medications, an approach that is already being used to find new treatments for Alzheimer's disease. Currently the main drugs for Alzheimer's disease improve brain function by prolonging the actions of a neurotransmitter called acetylcholine, a mechanism similar to the actions of some other contemporary drugs, such as the SSRIs. The identification of variants of APP, PS1, and PS2 in rare cases of Alzheimer's disease has helped focus attention on alternative drug targets. Called secretases, these are brain enzymes that play a part in the production of a toxic protein fragment called beta-amyloid, whose accumulation is also affected by the gene variants in a few different ways. Several drug companies are studying drugs that inactivate the secretases, which they hope to use to reduce beta-amyloid accumulation and thereby stop brain degeneration.

In addition to finding new drug targets, DNA studies may identify gene variants that distinguish people who benefit from available drugs, like SSRIs, from those who do not. Such distinctions could be caused by the particular variants that predispose an individual to a mental disorder and by others that determine how the drug affects the brain. The same DNA data may also reveal gene variants that influence individual vulnerability to certain side effects of drugs. All this genetic information will guide the selection of treatments for individual patients.

The DNA data will also help to redefine the boundaries between different mental illnesses, which often overlap. So, too, do the boundaries between the patterns of behavior we call normal and those we classify as disorders. Combining information about gene variants with studies of brain function, formal psychological tests, and a detailed life story will make it possible to replace crude diagnostic categories with a rich individual profile for each patient.

Fifty years from now, the reasons for a psychiatric consultation will not have changed. Some patients will be disabled by delusions of worthlessness or omnipotence, or by inexplicable attacks of panic, or by threatening voices echoing in their heads. Others will feel joyless, lifeless, pessimistic, chronically worried. Still others will just want to take stock of their lives.

But fifty years from now, everyone who visits a psychiatrist will bring with them a new source of information—a password providing access to their personal DNA file on the National Health Service computer. In that file will be the sequence of each of their genes, along with annotations calling attention to gene variants and combinations in that individual which influence vulnerability to a variety of disorders, and of others that influence the actions of drugs.

The initial consultation will take about an hour. A third of that time will be set aside for the completion of a formal questionnaire about personal development, family history, and specific symptoms. The rest will be an informal interchange. At the end of the session, the psychiatrist will offer an assessment, suggest some diagnostic tests, and request access to the patient's DNA file.

The request for such access will not seem untoward. The legislation that established a national repository of DNA files and ensured their privacy will also have set aside funds to publicize the benefits of making them available to appropriate professionals. Many people who consult psychiatrists will be eager to comply. This will be particularly true of those who come from families that are riddled with certain mental disorders; they may wish to get an assessment of their level of risk and find out if any preventive measures can be taken. Those who seek medication may opt to be guided by knowledge of their particular combination of gene variants.

Such guidance will be especially valuable, because there will be hundreds of medications to choose from. Some will be improved versions of those we have now, with more selective effects on neurotransmission. Others will have been developed on the basis of our new knowledge about brain functions. Still others will have followed from the identification of gene variants that increase the risk of mental disorders.

The availability of genetic information about mental disorders will not only change the diagnostic and therapeutic practices of psychiatrists but also augment psychiatry's contribution to the way we think about ourselves. In the first half of the twentieth century, psychiatry helped us realize that we are all heavily influenced by powerful innate passions and benefit from becoming aware of them. In the second half, it provided us with drugs for mitigating uncontrollable behaviors and showed how dependent we all are on simple brain chemicals, such as serotonin and dopamine. The identification of gene variants that influence behavioral differences will fill in some important details about each person's unique biology. Although it may prove difficult to interpret the significance of many of these gene variants, some of them will surely become useful tools for contemplating and constructing our individual life narratives.

❏

SAMUEL BARONDES, M.D., is the Jeanne and Sanford Robertson Professor and director of the Center for Neurobiology and Psychiatry at the University of California, San Francisco. He also serves as chair of the Board of Scientific Counselors of the National Institute of Mental Health. He is the author of *Molecules and Mental Illness* and *Mood Genes: Hunting for Origins of Mania and Depression* and is currently working on a book about psychiatric drugs.

❑

Brain Scans, Wearables, and
Brief Encounters

IN HIS AUTOBIOGRAPHICAL ESSAY *My Way to Hasidism*, the philosopher Martin Buber describes an encounter in Bukovina in 1910. He had just given a lecture and was relaxing at a local coffeehouse when a middle-aged man, identified only as "Mr. M.", approached him:

> "Doctor," he said, "I have a daughter." He paused; then he continued, "And I also have a young man for my daughter." Again a pause. "He is a student of law. He passed the examinations with distinction." He paused again, this time somewhat longer. . . . "Doctor," he asked, "is he a steady man?" I was surprised, but felt that I might not refuse him an answer. "Now, Mr. M.," I explained, "after what you have said, it can certainly be taken for granted that he is industrious and able." Still, he questioned further. "But Doctor," he said, "does he also have a good head?"
>
> "That is even more difficult to answer," I replied; "but at any rate he has not succeeded with industry alone, he must also have something in his head." Once again M.

paused; then he asked, clearly as a final question, "Doctor, should he now become a judge or a lawyer?"

"About that I can give you no information," I answered. "I do not know the young man, indeed, and even if I did know him, I should hardly be able to advise in this matter." But then M. regarded me with a glance of almost melancholy renunciation, half-complaining, half-understanding, and spoke in an indescribable tone, composed in equal parts of sorrow and humility: "Doctor, you do not want to say."

I am sometimes reminded of Buber's story during the two afternoons a week when I leave the orderly world of research and enter the wildly intimate world of psychotherapy. The sessions are filled with drama, humor, and pathos. My patients have so many questions:

"Should I leave my wife?"

"Am I a sex addict?"

"Why do I feel exhausted all the time?"

"Should I just stop talking to my brother, who always offends me?"

"How do I know that I won't be contaminated by touching CDs in record stores?"

"How can I be sure I won't harm my baby?"

It can make me want to run back to the lab and crunch some numbers and map some more prime brain real estate. Or it can make me feel humble and question the ways that our conversations may help.

At the dawn of the twenty-first century, we know so much more than we ever did about the circuitry of the brain, the computations of the mind, and the code embedded in the human genome. Although there is intense interest in translating this research into effective therapy, most of these advances lie in the future. At a time of giddy optimism in the neurosciences, it is a time of discontent in psychiatry and wary optimism in clinical psychology. If current trends continue, there will be few psychiatrists in practice fifty years from now. Fewer medical school students are choosing psychiatry as a specialty than at any time since 1929, and surveys show why: They perceive psychiatry as less helpful to patients, less intellectually challenging, less prestigious, and less financially rewarding than any other specialty.

But there is disenchantment among potential patients as well. The majority of people identified by community surveys as having psychiatric problems (defined by psychiatry's diagnostic reference, the Diagnostic and Statistical Manual of Mental Disorders—Fourth Edition) don't come in for therapy. Some people do not associate their symptoms with psychiatric disorders. Many say that they can handle their troubles on their own, or with the help of family or friends, or through prayer, rest, exercise, vitamins, pain relievers, a stiff drink. Some have no insurance; others feel ashamed, or worry about the stigma of a psychiatric diagnosis. But one survey points to a different problem: About half the respondents reported that they lacked faith in standard psychiatric treatments, such as medication and therapy. Psychotherapists have no shortage of clients, but the "worried well"—those who do not match any diagnosis—comprise a significant percentage of the people who seek their help.

This problem will become more acute. In their *World Mental Health: Problems and Priorities in Low-Income Countries* (1995), Robert Desjarlais and other members of the Harvard University Department of Social Medicine predict that the numbers of people with psychiatric disorders will increase substantially across the globe, in part simply because people are living longer, into the ages of risk for certain illnesses. By 2020, depression is predicted to be second only to ischaemic heart disease as the world's leading cause of disability. The increasing incidence of depression has been blamed on everything from social isolation and disruption of social roles to changes in global diet (such as low levels of omega-3 fatty acids) to changes in diagnostic criteria and methods of assessment to the spurious inflation of numbers of the depressed because of the hard sell of drugs like Prozac to psychiatrists and consumers.

I'm no futurologist, but I'll venture some predictions about psychotherapy in the twenty-first century. The current discontent is a good sign; it will usher in sweeping changes. I'll start with what I see as an inevitable shift of theoretical focus and close with some predictions about the way psychotherapy is likely to be transacted in the future. And yes, I will argue that people will still want to talk, even in the age of the user-friendly Prozacs of the future.

The Convergence of Knowledge

Here's an easy prediction: The finger-pointing at either nature or nurture—the tyranny of the single cause—will be thrown into the dustbin with history's other useless ideas. Psychiatric problems are as unlikely to be caused by a single gene or a single neurotransmitter (serotonin, dopamine, and so on) as they are to be caused by witnessing the primal scene or discovering that girls don't have penises. The origin and cause of most disorders is a complex interaction of genes and "the environment," a term that covers all nongenetic causes, including chance. It is likely that multiple genes acting as probabilistic risk factors will influence most psychiatric disorders.

A less obvious but inevitable development is that psychotherapists will no longer get away with thinking that the brain is irrelevant to what they do. In fifty years the study of the mind and the brain will not be divided among separate academic departments or professions, as it is now. The vicious territorial squabbles of the nineteenth century between psychiatry and neurology were settled by ceding the brain and its "organic" and "nervous" disorders to neurology and the mind and its "functional" and "mental" disorders to psychiatry. But of course all mental processes derive from computations in the brain, and research into the mind and the brain are part of a continuous terrain of knowledge.

For those psychoanalytic or humanist therapists who cannot imagine getting into bed with neuroscience, I offer for contemplation the human brain—an admittedly unaesthetic object when viewed with the naked eye but truly sublime in its beauty on closer inspection. This three-pound organ—packed with billions of neurons rivaling in their numbers the stars in our galaxy and equipped with up to two hundred thousand synaptic connections to other neurons—is the most complex structure in the universe. To scientists who use brain-imaging devices to watch the brain as it remembers, imagines, and desires, it is awe-inspiring. But the large question remains: How can this ebb and flow of blood, this intricate web of connections, become the experience of our feelings, the content of our thoughts? It is precisely this question that will occupy us in the next fifty years. As the geneticist François Jacob has written, "The century that is ending has been preoccupied with nucleic acids and proteins. The next one will concentrate on memory and desire. Will it be able to answer the questions they pose?"

What does brain science have to do with the practice of psychotherapy? The argument has been made (most forcefully by the neurobiologist Eric Kandel) that psychotherapy not only changes your mind, it changes your brain—literally. Effective therapy works in the same way and by the same mechanisms as any other form of intensive learning: It produces changes in gene expression that in turn change the strength of synaptic connections and produce structural changes that alter the pattern of interconnections between nerve cells in the brain. One might make the analogy to the training of a professional musician. The neurologist Alvaro Pascual-Leone has shown that the brains of professional musicians undergo functional and structural changes as they train, changes that can be documented by neuroimaging techniques. Pascual-Leone suggests further that even intensive mental rehearsals can bring such changes.

An increasing number of studies are comparing the effects of psychotherapy with those of drugs such as Prozac by looking at brain images before and after treatment. Such studies have been done for OCD (obsessive-compulsive disorder) and major depression. What they find is that when both forms of treatment are effective, they produce similar brain changes. The research suggests a common final pathway for complex psychological changes. In the future, simply by scanning the patient's brain, we may solve the seemingly unsolvable conundrum of how to judge whether a treatment is effective and when therapy should be terminated.

Freud Moves Out; Darwin Moves In

Influenced by Freudian theory, psychotherapists from the 1940s through the 1970s (and some of them even today) have taken for granted that psychiatric disorders are rooted in early childhood and that therapy must involve a detailed reconstruction of the distant past—an enormous task hindered by the limitations of memory. "I'm giving it one more year and then I'm going to Lourdes," Alvie Singer, Woody Allen's fictional stand-in, quips about his fifteen years of analysis in *Annie Hall*. Analysts defend the time requirements by citing the enormity of the task ("You can't dig the Suez Canal with a spoon"). But research has not confirmed early childhood trauma as the root cause of any psychiatric disorder; even posttraumatic stress disorder (PTSD) may arise from adult experience. While a microscopic analysis of a patient's early life might unearth some nuggets, current cognitive science suggests that memory is malleable. Forgetting, blocking, and occasional false memories are how the mind works, the price we pay for a memory system that generally serves us well.

In his 1997 article "Where Will Psychoanalysis Survive?" the psychiatrist Alan Stone concludes, "Psychoanalysis, both as a theory and as a practice, is an art form that belongs to the humanities and not to the sciences. It is closer to literature than to science." As Freud changes residence to the arts and humanities, Darwin will move into the behavioral sciences and medicine. In fifty years, Darwinian medicine will provide the framework for the field. The brain, like every other bodily organ, has been shaped by natural selection and has evolved mental modules that enhance reproductive fitness and help ensure survival. The practice of psychotherapy will be reoriented from a focus on disease to a focus on vulnerabilities, from symptoms to adaptive defenses, and from the exclusive focus on individual history to an examination of the shared aspects of human experience.

The prevalence of some disorders will be traced to the fitness benefits that the predisposing genes confer. Manic depression is an example: The energy, creativity, and charisma associated with mild mania may offer a fitness advantage to some people with the disorder, or to other people in whom the genes do not cause the disorder but have the beneficial effects. Other disorders will be understood as breakdowns in brain modules, such as (in schizophrenia) the system that normally distinguishes our own actions from the actions of others, or (in autism) the system that enables us to read other people's intentions and feelings. Certain symptoms will suggest design trade-offs prompted by mismatches between the present environment and the ancestral one, or simply exaggerated normal defenses. For example, the disorder known as general anxiety probably evolved as a defense against uncertain dangers, and phobias as a defense against specific ones, like bleeding or heights or venomous snakes. The flashbacks and recurrent memories of the PTSD sufferer may be torturous, but they arise because the mind finds it useful to remember life-threatening dangers in order to avoid such dangers in the future. Mild depression may serve the adaptive function of conserving resources in times of hardship, signaling others that help is needed, and allowing time for reassessment of goals. Mild depression may also be a sign of submission, when the individual cannot or does not wish to oppose the hierarchy.

From an evolutionary viewpoint, sadness, fear, anger,

disgust, shame, and guilt can be seen as adaptations and defenses; like coughs and calluses and physical pain, they serve useful functions. They can be overly noxious because they operate on what the University of Michigan psychiatrist Randolph Nesse calls the "smoke detector principle": It is better to issue a false alarm than to fail to detect a fire. Our environment is far safer than the ancestral environment—safer from infectious disease, malnutrition, predators, acts of nature. Perhaps, ironically, those now burdened by anxiety disorders would have been more fit in the environment of our distant past.

Evolutionary explanations raise important questions about treatment. If some symptoms—phobic fears, for example—are biological in origin, might that mean that they are untreatable? This is not the case: Fear of blood or fear of snakes and other animals can be eliminated with a few hours of exposure therapy. But while obsessive-compulsive disorder, depression, panic, and phobias are clearly disabling and benefit from treatment, mild anxiety and depression may be ultimately useful. Such painful states may help us to change course in life, question our own or others' decisions, make peace with friends or family, avoid danger. As Nesse has pointed out, it may be a mistake to medicate a society into being too fearless or too immune to sadness or loss.

In a related paradigm shift, the study of wellness will become as important as the study of illness. From fields as diverse as positive psychology to molecular genetics, we will see scientists investigating what protects people in times of adversity, what inoculates them against stress, and what genes or environment or temperament might be health-inducing. The field of "mental health" will no longer be a misnomer, since it will not exclusively study sickness.

From the Couch to the Wearable

At the heart of traditional psychodynamic therapy has been the patient-therapist relationship. Freud appeals to our literary imagination: The psychoanalytic dialog sweeps across the whole of a life, creating a satisfying and meaningful biography. As Lytton Strachey wrote in his preface to *Eminent Victorians,* biography is the "most delicate and humane of all the branches of the art." But even if HMOs hadn't come along, other treatments were bound to supplant the leisurely practice of psychoanalysis. Criticized as "an unidentified technique applied to unspecified problems with unpredictable outcomes," it was never meant for most people who are severely burdened by their psychiatric symptoms.

What is the future of talk therapy, and what will the conversation be about? Most of the time, psychotherapy will be problem-focused, brief, anchored in the present, relying on evidence-based techniques that have been tested for their effectiveness in treating specific problems, and guided by manuals. The provider will be a psychologist (who will have prescribing privileges) or a social worker— or, more rarely, a psychiatrist. The therapist's warmth and empathy will be important, but therapy will be less about the relationship and more about the exchange of information. The setting will be flexible and not always face-to-face; more and more, therapy will take place remotely, via the Internet (for education, self-diagnosis, and treatment monitoring); via Palm Pilots, which can offer instructions such as what to do during a panic attack; and by wearable devices. Psychotherapy will no longer be all talk and no action. In anticipation of long-term space travel, NASA is funding the creation of wearables that signal depression or anxiety or exhaustion. As the cosmonaut Valery Ryumin has remarked, "All the conditions necessary for murder are met if you shut two men in a cabin and leave them together for two months." Early-response tools can predict when an astronaut should stop working and be given treatment options ("Listen to your cognitive-behavior therapy tapes," or "Take an antidepressant," or "Take some time off and get retested in three hours.") Earthbound patients who want to exit treatment but monitor themselves for signs of relapse can also use such devices.

Computers of the future will be able to recognize our emotions. Wearable devices attached to clothing or jewelry or eyeglasses will gauge us by parameters we might never think to employ (our number of eye blinks or a furrow in our brow compared to our usual baseline), while implanted wearables can measure our internal workings. Roz Picard, of MIT's Media Laboratory, suggested in a 1998 interview in *The Atlantic Monthly* that emotion-sensing wearables will "pick up your literal and metaphorical smell. . . . [L]ike underwear, they will probably cease to be shared and will become truly personal computers." Well, perhaps they will be more like wearable therapists and less like panties.

Why will people bother with therapy when they can pop the drug cocktails of the future? If a medication can alleviate a stubborn, disabling symptom, most people will opt to take it. But most studies suggest that while drugs work for some people and therapy works for others, the combination often works best of all. Drugs tame symptoms, but therapy helps people to solve problems and learn solutions—not to mention that people are more likely to keep taking their medications if they are in therapy. For many people, psychotherapy alone will still be the best choice, affecting the brain in ways similar to drug therapy without the expense, side effects, or potential danger, and almost as quickly. And while drugs seem to work only as long as you take them, talk therapy promises the longer-lasting effects provided by learning.

On occasion, future therapy will bear closer resemblance to traditional therapy. Brief therapy with less face-to-face interaction will not work for some people—perhaps some of the same people who do not benefit from it now but require long-term treatment or treatment with no termination date. For them, the alliance with a therapist who offers empathy, careful listening, thoughtful conversation, a safe harbor, a relief from isolation, and a sense of shared purpose—a therapist who can help them compose a life—is what is healing. In the future world of brief encounters and nomadic wanderings, of brain scans and brain stimulators, there will still be ongoing conversations that change the way we see the world.

❑

NANCY ETCOFF is a member of the Harvard University Faculty of Medicine, the Massachusetts General Hospital Psychiatry staff, and the Harvard Mind/Brain/Behavior Initiative. Dr. Etcoff's research on the perception of beauty, emotion, and human faces has been published in *Nature, Cognition, Neuron,* and other scientific journals, has been cited frequently in the popular press, and has won numerous awards. She is the author of *Survival of the Prettiest: The Science of Beauty.*

PAUL W. EWALD

❑

Mastering Disease

WHAT CAUSES DISEASE? The question is so basic, modern technology so sophisticated, and promises from experts so confidently advanced that an outsider might presume that the causes of diseases are well understood. They are not. Medicine is still struggling to grasp the causes of many devastating chronic disorders: heart attack, stroke, Alzheimer's disease, schizophrenia, cancer, and diabetes. The quality of our lives over the next fifty years will depend on how well these chronic diseases are controlled.

As is the case with all biological phenomena, causation can be considered in its mechanistic sense ("What are the agents that bring about the disease?") and in its evolutionary sense ("What are the selective pressures that result in the disease?") In the mechanistic sense, a consensus on causation exists for only about half the diseases listed in current medical textbooks. The mechanistic causes can be grouped into three categories: genetic, parasitic (including infection), and nonparasitic environmental influences such as radiation and too much or too little of particular chemicals.

Most experts in the health sciences advocate a building-block approach to the problem of causation. They try to understand the workings of disease at the cellular and biochemical levels, in hopes that solutions will eventually emerge. This is an appealing idea, but it hasn't yet generated any of the great successes of medicine—great success being decisive solutions rather than patch jobs. What Lewis Thomas wrote thirty years ago in his essay "The Technology of Medicine" is still true today: The vast majority of medical practice, from organ transplants and bypass surgery to most cancer therapy, is devoted to patch-job solutions or supportive care of marginal value. Experts respond that today's diseases are less tractable than those of the past and patch jobs may be the best we can expect. Yet the record suggests otherwise. Decisive solutions continue to be found for diseases recently considered intractable. Generally such solutions have come from understanding infectious causes; for example, over the past two decades countless thousands of cases of liver cancer and hepatitis have been prevented by screening blood supplies for hepatitis B and C viruses and by the hepatitis B vaccine, and peptic ulcers and stomach cancer can now be cured with antibiotics.

Even among infectious diseases, however, the fundamental achievements have occurred more through the testing of deductive leaps than by building-block induction. Edward Jenner had no understanding of viruses when he founded modern vaccination more than two hundred years ago. John Snow and Ignaz Semmelweis had never seen a bacterium when they founded modern epidemiology a half century later by showing how infectious diseases were transmitted and how they could be prevented by blocking that transmission. Nor had Joseph Lister when he demonstrated the effectiveness of the sterilization of surgical equipment. Paul Ehrlich and Alexander Fleming knew virtually nothing about the mechanisms of chemical inhibition of bacterial growth when they established the concept and practice of antibiotic therapy. This history emphasizes the value of looking beyond the building-block approach that guides modern medical research to consider conceptual insights that might generate clusters of decisive solutions. One such approach is to assess disease causation from an evolutionary perspective.

Evolutionary explanations are based ultimately on genetic causation. One might therefore presume that evolutionary medicine would explain human diseases largely in the context of human genetics. Evolutionary medicine does in fact offer such explanations. Chronic diseases, for example, have been explained by senescence theory: The body falls apart because natural selection favors characteristics that benefit individuals at younger ages but impose a cost when they are older and natural selection less potent. Alternatively, chronic diseases may result from a mismatch between modern environments and the environments that were mostly responsible for shaping human genes. A third major alternative is infectious causation: Chronic diseases may be a consequence of infectious agents that cryptically cause tissue damage, which eventually manifests itself in such serious diseases as heart attack, cancer, or Alzheimer's. This last alternative does not negate the value of casting evolutionary explanations ultimately in the context of genes, but emphasizes that we must not restrict hypotheses about causation of chronic diseases to human genes. We must also consider genes of parasites.

My colleague Gregory Cochran and I have emphasized that for the common damaging chronic diseases the evidence considered in light of evolutionary principles implicates infection. If these diseases were due to defective alleles, mutation rates would generally be low to maintain the diseases at their observed frequencies. The patterns of human survival and disease occurrence do not fit the senescence trade-off model, and though new environmental factors seem plausible in principle and important in some cases, such as smoking-induced lung cancer, the proposed noninfectious environmental causes are generally insufficient by themselves to explain the patterns of disease. Adding infectious causation into the mix can best explain the documented epidemiological patterns, and do so in accordance with evolutionary principles, because the genetic conflicts of interest between pathogen and host can perpetuate states of disease indefinitely over time. In the short run, a genetic vulnerablity to a particular pathogen would tend to be disfavored by natural selection much as a bad gene would. But unlike bad genes, genetic vulnerabilities to infectious disease are continually changing over the long run in a coevolutionary chess game. The selection against genetic vulnerability to a pathogen leaves as a legacy a host population that has countermeasures against the pathogen. These countermeasures in turn favor the evolution of counter-countermeasures in the pathogen population, and so on indefinitely.

Most problems in evolutionary biology are resolved slowly, because the problems are complex, testing is difficult, and funding is scarce. In contrast, the causation of chronic diseases will probably be resolved relatively rapidly—not because researchers in the traditionally demarcated disciplines of the health sciences will eagerly consider their subject in the light of evolution but because they are now debating and avidly researching infectious causation of chronic diseases. Given the implications of evolutionary theory, the march of medical research, and the accumulated evidence, I expect that the common and highly damaging chronic diseases—atherosclerosis, diabetes, Alzheimer's disease, most cancers, and most fertility problems—will, in the next fifty years, be accepted as caused by infection.

But this prediction is rather spineless; by the time it can be evaluated I probably won't be around to be criticized for being wrong. If the evolutionary approach to disease causation is of real value, we should be able to take on the riskier challenge of specifying diseases for which infectious causation will be accepted during the next few decades. Table 1 lists some of the most important chronic diseases, along with the years by which I predict they will be accepted as being caused by infection. To make these predictions testable, I suggest the following criterion for "acceptance": At least three-quarters of the medical texts published in the preceding five years must refer to the disease as being caused by infection.

To presume that acceptance of infectious causation depends mainly on the quality of the evidence is to place too much weight on objective evaluation. Steadily accumulating evidence does help, as long as the appropriate range of hypotheses is being tested. But we probably will never get the kind of evidence for infectious causation that holdouts demand, because absolute proof of infectious causation will probably be unobtainable for most of the remaining chronic diseases caused by infection. Over the past quarter century, for example, the medical community has accepted human herpes virus 8 as the cause of Kaposi's sarcoma and HTLV-1 as the cause of adult T-cell leukemia. But the same quality of correlational evidence would not be accepted for atherosclerosis, chiefly because the many

Table 1. Predicted years for acceptance of infectious causation of various chronic diseases.

Disease	Candidate pathogens	Year by which infectious causation will be accepted
multiple sclerosis	*Chlamydia pneumoniae*, human herpes 6	2010
type 2 diabetes	hepatitis C virus (minor cause)	2010
type 2 diabetes	unknown (major causes)	2025
head and neck cancers	human papillomavirus	2010
nasopharyngeal cancer	Epstein Barr virus (EBV)	2010
childhood leukemia	unknown	2015
breast cancer	mouse mammary tumor virus, EBV	2015
atherosclerosis	*C. pneumoniae, Porphyromonas gingivalis, cytomegalovirus, Actinobacillus actinomycetocomitans*	2015
Alzheimer's	herpes simplex 1, *C. pneumoniae*	2015
schizophrenia	*T. gondii*, herpes simplex 2, endogenous retrovirus, borna disease virus (BDV)	2020
bipolar depression	BDV	2025
prostate cancer	unknown retrovirus	2025

decades of research and expert opinion on that disease have generated a great diversity of vested interests that create a drag on the pace of acceptance. It takes time to be freed from this drag. Here the insights of Charles Darwin, Max Planck, and Thomas Kuhn apply: A sufficient proportion of the old guard will have to retire or expire, and a sufficient number of young people entering the arena without these vested interests must mature into positions of influence, to tip the balance of expert opinion. My guess is that this point will have been reached by 2015 for atherosclerosis and Alzheimer's disease. There is some dark irony here: The causal organisms contribute to this process by dethroning those who dismiss them.

The predictions in this table use transitions of the past as a guide but do not simply assume that the time between the first appearance of convincing evidence and acceptance is constant. Another consideration is the size of the leap from what is already accepted as caused by infection. I expect that infectious causation of some cancers of the head and neck will be accepted sooner than infectious causation of atherosclerosis, even though the evidence for the former is now less complete. The main reason is precedence. The candidate pathogen, human papillomavirus, has already been accepted as a cause of cervical cancer; moreover, balanced consideration of causal explanations for these cancers (like those of adult T-cell leukemia and Kaposi's sarcoma) are not as burdened by the drag of strong vested interests. Similarly, if and when *Chlamydia pneumoniae* is accepted as a cause of any one of the major chronic diseases in which it is implicated—atherosclerosis, stroke, Alzheimer's disease, and multiple sclerosis—resistance to its acceptance as a cause of the remaining diseases should fall like dominoes. In this case, however, the commonality is not tissue type but a common genetic predisposition—the epsilon-4 allele, which has been associated with increased vulnerability to C. *pneumoniae*.

The general opposition to infectious causation of cancer seems particularly ingrained. When considered on the basis of the actual evidence, this resistance is surprising, because medical researchers already accept infectious causes for about 15 to 20 percent of human cancers (up from less than 1 percent twenty-five years ago) and there are so few examples of human cancers for which infection can be ruled out (less than 5 percent). The war over infectious causation of human cancer has been going on since about 1910, when Peyton Rous showed that viruses could cause cancer in chickens. The most ardent antiinfection advocates declared victory in the late 1970s, when contributions by oncogenes and noninfectious carcinogens were confirmed and a feasible mechanism was thus obtained. They have continued to declare victory, while retreating incrementally year by year. Their slip of logic involves mistaking evidence in favor of one contributing mechanism as evidence *against* contributions by another (mutually compatible) mechanism. All three categories of disease causation often act in concert.

As with atherosclerosis and cancer, acceptance of infectious causation of severe mental illnesses, such as schizophrenia and bipolar disorder, are bound to be delayed because of vested interests of the experts. Added to this drag is the specificity of mental illnesses to humans, which makes experimental demonstration of infectious causation difficult. How does one know when a mouse is hallucinating, paranoid, depressed, or manic? To counter this argument, one might point to the early recognition of syphilitic insanity. But infectious causation of syphilitic insanity was easily accepted because researchers had already linked syphilitic insanity to syphilis. Once syphilis was accepted as caused by infection, the experts were primed to accept the idea that syphilitic insanity was caused by infection. Syphilitic insanity was then conveniently sequestered in a separate domain of causation within the realm of mental illness. Today's mental illnesses with unknown causes have not been linked to clearly identifiable acute illnesses; in this sense, they are more like adult T-cell leukemia and Kaposi's sarcoma than they are like syphilitic insanity. Why then am I optimistic about imminent acceptance of the infectious causation of schizophrenia? Epidemiological patterns strongly implicate infectious causation; for example, schizophrenics tend to be born in late winter or early spring, a clue that schizophrenia may result from prenatal or neonatal infections that peak during winter or early spring. The discovery of significant associations with candidate pathogens also implicates infectious causation. A recent study, for example, showed a positive reaction for the brain parasite *Toxoplasma gondii* in 42 percent of schizophrenic patients but only about 11 percent of healthy controls. Infectious causation of schizophrenia is now being treated seriously by experts who would have dismissed the idea a decade ago, and several labs are actively investigating the possibility. All of this suggests to me that the acceptance of infectious causation of schizophrenia will result largely from the retire-and-expire factor, which should tip the balance of opinion within the next ten to fifteen years.

Over the next five decades, the development of vaccines against infection-induced chronic diseases should follow closely our discovery of the causal agents. When a pathogen is not particularly mutable, vaccines can be particularly effective. We can therefore expect DNA viruses, such as human papillomaviruses, to be more controllable by vaccines than RNA viruses, such as HIV. This future mastery of disease requires no fundamentally new technological advances. We already know that the right vaccines will prevent disease and the right anti-infective agents can cure diseases and prevent infections from progressing to damaging states. We already know that breaking the chain of transmission can prevent infection in an individual and sometimes do so for an entire population. And the medical track record for developing such solutions for newly recognized infectious agents has been very good.

For infectious diseases, evolutionary causation pertains largely to the evolution of virulence: What causes some parasites to evolve to benignity, lethality, or somewhere in between? The answers to this question hold promise of a third major approach to controlling infectious disease; that is, in addition to curing infection and controlling the spread of infection, we should be able to control the evolution of the pathogen, thereby transforming life-threatening adversaries into benign cohabitants. For each of the major categories of diseases, theory and the available evidence indicate that there is at least one kind of intervention that could accomplish this goal. During the next quarter century, I expect that the first of these evolutionary success stories will be demonstrated, probably through the clever use of vaccines or by inhibiting waterborne transmission of diarrheal pathogens in poor countries. Also hanging in the balance is control of antibiotic resistance: The more benign the pathogens, the less prevalent the use of antibiotics and therefore the less evolution of antibiotic resistance. If we can control the evolution of virulence we should also therefore be able to control the evolution of antibiotic resistance.

If the history of medicine is any indication, the first success story will encourage the same approach to other diseases—as was the case with vaccines, antibiotics, and hygienic improvements. However, because the timing of the first success is difficult to predict, the timing of this chain reaction is difficult to predict. By mid-century we will probably have enacted the evolutionary transition from virulence to benignity for a few diseases, such as the damaging diarrheal diseases, but we will still be in the midst of the testing process for most of the rest.

What *won't* happen in the next fifty years? We are bound to run up against some agents that stymie our efforts at decisive control. For example, unless some fundamentally new measure is brought to bear on HIV, it will still be a problem suppressed but not solved by the patch jobs—that is, the antivirals and vaccines that constrain but do not decisively control the virus. The genetic flexibility of HIV is to blame for this dreary outlook: It readily evolves resistance to antivirals and can be expected to escape from the straitjackets of vaccination. Clever use of combinations of antivirals may give medicine an advantage, as it has over the past several years, but will not provide a cure. The best hope is to buy time—a decade or so of AIDS-free life from antivirals, perhaps a decade or so from therapeutic vaccines administered soon after infection, and perhaps a decade or so by controlling the evolution of virulence. Those decades add up but still will not provide the kind of decisive control that medicine has imposed on diseases such as diphtheria, smallpox, whooping cough, measles, and polio—especially considering the side effects associated with prolonged use of antiviral compounds.

Despite current worries about horrific new emerging diseases sweeping the globe, we can be reasonably certain that over the next half century the number of devastating new plagues like the AIDS pandemic will be zero or one—probably zero. For well over a century there has been a global mixing of people, ever more rapidly and from ever more remote regions. This mixing must by now have exposed us to pathogens from virtually every part of the world. Yet in spite of all this mixing, there has (as far as is known) been only one new pathogen that has spread from a geographically restricted area to cause a devastating global pandemic: HIV-1. Using the AIDS pandemic as an example, scientists and laymen alike have warned that other pathogens might pose similar threats. The familiar candidates include exotic hemorrhagic viruses such as Ebola and Marburg; the mosquito-borne West Nile virus; hantavirus (the cause of the recent "Four Corners" disease in the American Southwest); and the prion that causes mad cow disease and its human counterpart, new variant Creutzfeld-Jakob. But these worries are misplaced. These pathogens, though terrible for those whom they sicken, do not pose the threat of global disaster. They simply do not have the right characteristics. West Nile virus cannot be transmitted from humans to mosquitoes. Ebola virus is not sufficiently durable in the external environment to maintain cycles of lethal human infection. The prions are transmissible between people only by unusual and controllable activities, such as cannibalism and cornea transplants.

The real threat comes from those pathogens that are already among us and killing us or may evolve increased ability to do so. If the current state of evidence implicating infectious causation is on target, we are already dying of terrible global pandemics of heart attack, stroke, Alzheimer's disease, and cancer—pandemics caused by infectious agents that are lethal but overlooked. We have been worrying about a few stray cats, while we are being stalked by leopards. Over the next half century, we will finally recognize the leopards, and we will take measures to keep at least some of them at bay.

❑

PAUL W. EWALD is a professor of biology at Amherst College and a specialist in evolutionary medicine, a discipline he helped to found and on which he has lectured extensively at college campuses, seminars, and symposia around the world. He is the author of *Evolution of Infectious Disease* (acknowledged as the watershed event in the emergence of that discipline) and *Plague Time: How Stealth Infections Are Causing Cancers, Heart Disease, and Other Deadly Ailments.*